T0358738

ACTIVITY THEORY IN FORMAL AND INFORMAL SCIENCE EDUCATION

CULTURAL AND HISTORICAL PERSPECTIVES ON SCIENCE EDUCATION: RESEARCH DIALOGS

Series editors

Kenneth Tobin, *The Graduate Center, City University of New York, USA*
Catherine Milne, *Steinhardt School o Culture, Education, and Human Development, New York University*

Scope

Research dialogs consists of books written for undergraduate and graduate students of science education, teachers, parents, policy makers, and the public at large. Research dialogs bridge theory, research, and the practice of science education. Books in the series focus on what we know about key topics in science education – including, teaching, connecting the learning of science to the culture of students, emotions and the learning of science, labs, field trips, involving parents, science and everyday life, scientific literacy, including the latest technologies to facilitate science learning, expanding the roles of students, after school programs, museums and science, doing dissections, etc.

Activity Theory in Formal and Informal Science Education

Edited by

Katerina Plakitsi
University of Ioannina, Greece

SENSE PUBLISHERS
ROTTERDAM / BOSTON / TAIPEI

A C.I.P. record for this book is available from the Library of Congress.

ISBN 978-94-6091-315-0 (paperback)
ISBN 978-94-6091-316-7 (hardback)
ISBN 978-94-6091-317-4 (e-book)

Published by: Sense Publishers,
P.O. Box 21858, 3001 AW Rotterdam, The Netherlands
https://www.sensepublishers.com

Printed on acid-free paper

TABLE OF CONTENTS

1. Activity Theory in Formal and Informal Science Education:
   The ATFISE Project 1
   Katerina Plakitsi

2. Cultural-Historical Activity Theory (CHAT) Framework and Science
   Education in the Positivistic Tradition: Towards a New Methodology? 17
   Katerina Plakitsi

3. Teaching Science in Science Museums and Science Centers:
   Towards a New Pedagogy? 27
   Katerina Plakitsi

4. Rethinking the Role of Information and Communication Technologies
   (ICT) in Science Education 57
   Katerina Plakitsi

5. Chat in Developing New Environmental Science Curricula, School
   Textbooks, and Web-Based Applications for the First Grades 83
   *Katerina Plakitsi, Eleni Kolokouri, Eftychia Nanni, Efthymis Stamoulis
   and Xarikleia Theodoraki*

6. Activity Theory, History and Philosophy of Science, and ICT
   Technologies in Science Teaching Applications 111
   Efthymis Stamoulis and Katerina Plakitsi

7. University Science Teaching Programs: What's New in Lab Activities
   from a Chat Context? The Case of Magnetism 159
   Xarikleia Theodoraki and Katerina Plakitsi

8. A Cultural Historical Scene of Natural Sciences for Early Learners:
   A Chat Scene 197
   Eleni Kolokouri and Katerina Plakitsi

9. Biology Education in Elementary Schools: How do Students Learn
   about Plant Functions? 229
   Eftychia Nanni and Katerina Plakitsi

List of Contributors 253

v

KATERINA PLAKITSI

1. ACTIVITY THEORY IN FORMAL AND INFORMAL SCIENCE EDUCATION

The ATFISE Project

INTRODUCTION

This book aims to contribute to an emergent agenda for cultural historical activity theory (CHAT) and science education in Europe. It especially focuses on the application of activity theory in formal and informal science education. This focus leads to rethinking scientific literacy (Roth & Lee, 2004), as well as to rethinking the role of information and communication technologies (van Eijck & Roth, 2007; Kaptelinin & Nardi, 2006). Recently, many European science curricula have been reformed, but by interpreting evaluation reports of the Programme for International Students Assessment (PISA, 2006, 2009)[1] we see that we still have to do a lot in order to achieve the aim of "real" scientific literacy.

CHAT is considered a subcategory of sociocultural theory, and this issue will be analytically described in Chapter 2. A science education enriched and interpreted by CHAT could be situated in the current sociocultural context. During recent decades many scholars in the United States, Canada, Australia, and Europe have developed theoretical documentation and research methods on CHAT. Some important academic journals in science education research, such as *Science Education, Research in Science Education,* and *Journal of Research in Science Teaching,* increasingly include cultural studies of science education. The journal *Cultural Studies of Science Education* is totally oriented to this emerging research field. In this journal many senior and new authors publish work devoted to the cultural interpretation of science education practices and activities.

Among European science education policies, however, this emergent agenda remains isolated, although "learning communities," "potentials for learning," and "quality in science education research" are major topics in recent European journals, conferences, and books.[2] European science education scholars are underrepresented in this research area. For example, during the European Science Education Research Association (ESERA) conferences, few symposia were dedicated to cultural studies of science education (CSSE). Moreover, the average number of sociocultural articles in the leading European science education journal, *International Journal of Science Education* is low. We need more concerted work on major sociocultural and cultural-historical issues. Until now the discourse has been limited primarily to language, globalization, and immigration. European citizens differ from those in third-world countries, while science approaches in European countries may differ

K. Plakitsi (ed.), Activity Theory in Formal and Informal Science Education, 1–15.

significantly from those in Canada, the United States, and Australia. Furthermore, many types of science, for example science of western civilizations, personal science and indigenous science, can occur simultaneously in a learning community.

The traditional dualistic framework does not help us understand current complex social interactions. More than ever before, there is a need for an approach that can dialectically link the individual with social structure. From its very beginnings, the Cultural-Historical Theory of Activity (CHAT) considered this task as a priority (Engeström, 1999). Activity theory has its origins in classic German philosophy (from Kant to Hegel), in the writings of Marx and Engels, and in the Soviet Russian cultural-historical psychology of Vygotsky, Leont'ev, and Luria. Today activity theory is becoming truly international and multidisciplinary. This process entails the discovery of new and old related approaches, discussion partners, and allies, ranging from American pragmatism and Wittgenstein to ethnomethodology and theories of self-organizing systems (Engeström, 1999, p. 20). Activity theory is a framework or descriptive tool (Nardi, 1996) that provides "a unified account of Vygotsky's proposals on the nature and development of human behaviour" (Lantolf, 2006, p. 8).

Two of CHAT's most important contributions concern mediation and changes in human behavior. The first idea is that mediation with tools is not merely an idea. It is an idea that breaks down the Cartesian walls that isolate the individual mind from culture and society. The tools are both mental and physical. Examples of mental tools are the ability to measure, language (langue), and even some historical scientific experiments which changed our world. Examples of physical tools are magnifying glasses, simple balances, a textbook, operations on a PC, a social robot, or language (parole). Tools take part in the transformation of the object into an outcome, which can be desired or unexpected. They can enable or constrain activity.

The second important idea is that humans can control their own behavior—not "from the inside," based on biological urges, but "from the outside," using and creating artifacts.

Describing in brief the components of an activity represented in Figure 1, we mention subject, object, tools, rules, community, division of labor, and outcomes.

The subject of an activity system is the individual or group whose viewpoint is adopted.

An object "refers to the 'raw material' or 'problem space' at which the activity is directed and which is molded or transformed into outcomes with the help of physical and symbolic, external and internal *tools*" (Engeström, 1993, p. 67, italics in the original). It precedes and motivates activity.

The interaction between the subject and the object is mediated by the *tools*, but it is simultaneously influenced by the rules, the community, and the division of labor.

The rules are explicit and implicit norms that regulate actions and interactions within the system (Engeström, 1993; Kuutti, 1996).

Community refers to participants in an activity system who share the same object.

The division of labor involves the division of tasks and roles among members of the community and the divisions of power and status.

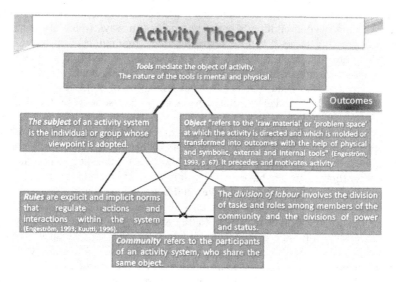

Figure 1. Components of the activity system (Engeström, 1987).

Apart from the basic triangle of CHAT, many prominent socioculturalists have supported some major trends of the theory. We focus on the concept of participation (Roth & Lee, 2004). Science education as participation in the community can work as a syllabus for teachers/researchers in science education who are rethinking the scope of scientific literacy. The core tendency is to construct theoretical assertions from an example or a case study. Some may consider this approach to be methodo-logically problematic, but we oppose this view, because each specific situation can contribute to a bottom-up approach to rethinking science education in a sociocultural context. We also oppose the formation of the theoretical assertions following a top-down approach, for example, from general pedagogical principles to everyday practices. We believe that it cannot help practitioners apply CHAT in their everyday settings because of the gap between general principles and practice.

Furthermore, a very recent study describes children's development as participation in everyday practices across different institutions (Hedegaard & Fleer, 2010). Institutions can either be the home or the school that most children share. Apart from the differences, there is a common core framed by societal conditions. Two theories can be combined in this approach: (1) Vygotsky's theory (1998) of the social situation of development and (2) Hedegaard's (2009) theory of development as the child participates within and across several institutions. The processes within and across those institutions have to be considered dialectically. This leads to the necessity for a new epistemology,[3] which is multiculturalism. Multiculturalism fits Hedegaard's psychological theory, as it legitimizes the different institutions as frameworks of knowledge acquisition and behavioral change. In this dialectical process "in which a transition from one stage to another is accomplished not as an evolving process but as a revolutionary process" (Vygotsky, 1998, p. 193), Fleer and Hedegaard (2010) invite teachers to project-based learning beyond children's current capacities in ways which are connected with their growing sense of themselves within their communities or institutions (p. 150). Consequently, teachers need to do a context analysis and study the evolution of children's conceptualization of scientific issues. Teachers must locate the points of crisis, always taking into account the social situation of child development. We can assert that Vygotskian revolutionary theory corresponds to the Kuhnian revolutionary epistemology about science. We also have to study not only changes in the child and changes in the environment, but changes in the child's relationship with the environment (Kravtsova, 2006). Danish and Australian case studies in Fleer and Hedegaard's (2010) work illustrate the conflicts within the child – inner conflicts – which have a major influence on the child's behavior, on relationships with the teacher and other children, and on his or her knowledge acquisition process. There is a great deal of literature on this topic; we only mention the argument on development to the extent to which "development can be understood only in light of the cultural practices and circumstances of their communities – which also change" (Rogoff, 2003, pp. 3-4). According to Rogoff (2003), development can be viewed as a transformation of participation in cultural activities, through which individuals change, thereby changing the communities within which they live.

In Hedegaard's work, the concept of institutional practice is the connecting link between Rogoff and Vygotsky's points of view and a step forward to the relevant discussion. According to the latest discussion, we are challenged to see that development takes place when a child participates in practices through different institutions. Figure 2 illustrates Hedegaard's model of development, which is strongly related to the Vygotskian tradition of societal development. We think that this visualization provided by Marianne Hedegaard can help teachers and researchers better understand the relationships within societal, institutional and individual participation in childen's development. Moreover, we can expand this approach, grounding our projects described in Chapters 7 and 8 on the dialectical participation of formal and informal science education.

Combining Vygotsky (1998) and Hedegaard (2009), we should not fear crisis but rather should see crisis as a dynamic context for development. We think that this conception of crisis can lead us to the opposite of cognitive conflict in the Piagetian tradition. Cognitive conflict, however, is considered an inner procedure, and its solution can be mediated by the teacher. It is therefore oriented more to the inner child and his or her cognitive domain and does not take into account the total pedagogical and societal environment. Researchers on social constructivism tried to take into account the societal factor in child development, but they remained anchored to the individuality and did not address the gap between theory and praxis.

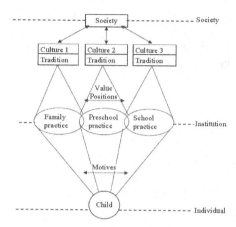

Figure 2. A model of child's learning and development through participation in institutional practice (Hedegaard, 2009, p. 73).

Overall, we propose that the framework provided by activity theorists is a coherent theoretical framework which establishes science education as participation in the community. Moreover, CHAT addresses the gap between theory and praxis. Also, it could achieve the scope of interdisciplinary science education in multicultural Europe. Consequently, a new mentality, which sees situated science education as part of society, has emerged. This could reform science education from its core, while lifelong learning activities take place in and for the community and for individuals as well.

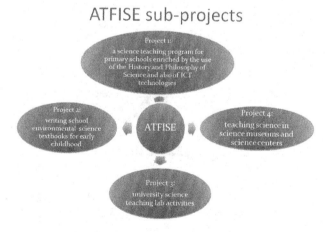

Figure 3. ATFISE subprojects.

We tested our proposal in four different settings: (1) a science teaching program for primary schools enriched by using the History and Philosophy of Science and

ICT technologies, (2) school environmental science textbooks for early child-hood, (3) university science teaching lab activities, and (4) science museums and science centers. The ATFISE subprojects are represented in Figure 3.

In continuation, there is an introduction to each ATFISE subproject.

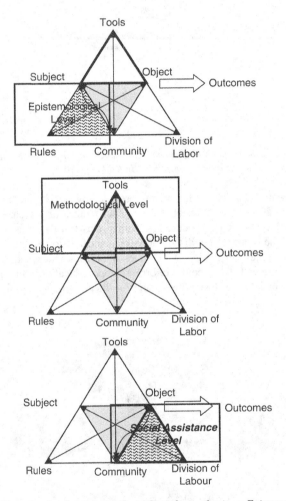

Figure 4. Three levels of activity analysis. Specific emphasis on Epistemological, methodological, and societal interactional levels.

In Chapter 6, ATFISE subproject 1 uses CHAT to analyze and then design new ICT learning environments enriched by the History and Philosophy of Science, which are the prominent cultural mediation tools. It focuses on parts of Engeström's triangle[4] focusing either on the epistemological level (rules-subject-community), or the methodological level (tools, subject, object), and|or the societal interactions level (division of labor, object, community) (Figure 4).

In Figure 5 we see a web page for ATFISE subproject 1. A welcome page for children, introduces an interactive lesson, in which we used several teaching strategies enriched by the history of science.

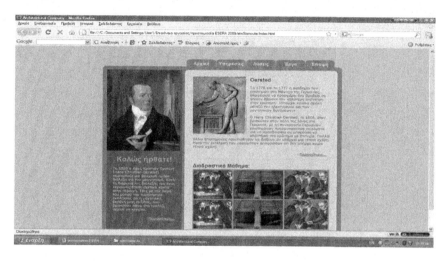

Figure 5. A web page for Sub-Project 1 (photo from editor's/author's archives).

To develop teaching activities, we employ Engeström's (1987) conceptual tool of the expansive cycles. In Figure 6 we use Engeström's (1999) descriptions of the "ideal-typical sequence of learning actions." In Chapter 6, we expand this idea by using expansive cycle as a tool for designing activities for primary science education.

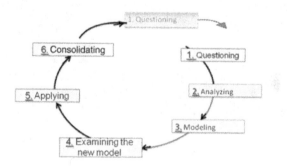

Figure 6. Expansive cycle (Engeström, 1999, p. 383).

ATFISE subproject 2 (see Chapter 5 in this volume) was concerned with the development of school science curricula and textbooks for the first grade, as well as two environmental software programs for elementary schools. We tried to develop those materials in the sociocultural context, and for this reason we used salient cultural tools within and across multiple authentic learning environments (Figure 7).

Figure 7. School science textbooks and software for first grade
(photos from the editor's archives).

ATFSE subproject 3 (parts 1 and 2) was presented at the ESERA biennial meetings in 2009 and 2011 and is concerned with applying activity theory in university lab lessons, as well as using cartoons in teaching science in the early grades. The conspicuous cultural tools are the cartoon stories we wrote and projected in the classroom. While university students were working in the science lab, we recorded their dialogue exchanges and experimental practices and then we analyzed the group interactions according to Mwanza and Engeström's eight-step model (2003).

ATFISE Project 3.1

Activity Theory and learning in Science Education laboratory lessons. The case of magnetism.

ATFISE Project 3.2

Scientific Literacy and Nature of Science in Early Grades using Cartoons

Figure 8. Project 3.1: Students/future teachers in early childhood experiment with magnets in science lab(photos from the editor's archives). Project 3.2 Future teachers use popular cartoons for teaching sinking and floating things (http://www.nick.com/games/spongebob-game-builder/).

ATFISE subproject 4 connects formal and nonformal astronomical learning. This project refers to an astronomy education program for preprimary and primary school students, which aims to develop a new science curriculum within museum education programs and introduces methodological tools from CHAT.

The placement of a museum piece, such as a mobile inflatable planetarium, inside a typical school allows us to explore interactions between formal and nonformal education, to experiment with new teaching processes using activity theory, and to track similarities and differences between our case and the usual situation, when the planetarium is a permanent installation out of the school.

*Figure 9. Children in and outside the mobile planetarium
(photos from the editor's archives).*

Furthermore, we organized a Lifelong Learning Program (Erasmus Intensive Program) entitled LIGHT with the participation of seven European University Departments related to Cultural Studies of Science Education (http://erasmus-ip.uoi.gr).

During this interdisciplinary and multicultural project, university students were moving, for example, from the class to the lab and then outside to observe the night sky and then to the video seminar room.

In Figure 10 there is an example from the triangle analysis of the video seminar they conducted on the solar eclipse.

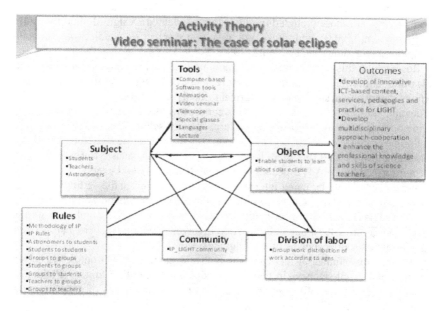

Figure 10. Triangular analysis of a video seminar on the total eclipse physical phenomenon (editor's archive).

The subjects were students, teachers, and astronomers who were involved in cogenerative dialogues (Roth & Tobin, 2004). The intensive program has rules established by the European Committee (e.g., to work at least 8 working hours per day), but the observation of the night sky had to be done after dusk, so all groups had to interact via other means and in a specific place outside. The community was strongly multicultural, with students and teachers from seven European countries, many religions (e.g., Christians and Muslims) and races (black and white), people from northern Europe and people from the Mediterranean. There were many tools, such as computer-based software tools, animations, video seminars, telescopes, special glasses, languages, and lectures. The goal was to enable students to learn about the total eclipse of the moon. The outcomes moved further, as the participants developed innovative ICT-based content, services, pedagogies, and practice about the properties of light. Finally, the development of such a multidisciplinary approach emphasized cooperation, and the enhancement of the professional knowledge and skills of science teachers.

In all mentioned subprojects, the main characteristics of the applied activities were the cultural profile of the learning environments, the cultural-historical references, and the cultural-historical means and methods of analysis. Our study belongs to the third generation of activity theory, which is concerned with understanding and modeling networks of activity systems.

The theoretical and methodological framework of analysis was the developmental approach of Yrjö Engeström (1987, 2005). Key elements of our methodology are those included in the Activity-Oriented Design Method (Mwanza & Engeström, 2003), and these are related to scientific studies on "human-computer interaction" (HCI) (Kuutti, 1996; Nardi, 1996).

Research on this interaction, using a nondualistic basis as an inseparable part of a learning-and-doing system, is much more than a challenge. We are going to adapt activity theory as a designing tool, in formal (schools) and informal (museums) science education sites, by using e-settings. This concept would advance the diffusion of a common European science learning culture. Modern schools and science museums in Europe organize many indoor and outdoor scientific activities based especially on e-learning technologies, but there is no common European science learning environment informed by CHAT, especially for young learners (5 to 9 years).

We collected data by using interviews, video-recordings and e-settings. All data are concerned with how science education is progressing in schools and labs (formal) and museums (informal). Specifically, as has been proposed in a number of studies (Roth and Tobin, 2005), our field research involves children, teachers, parents, and non-formal educators such as museum guides, etc

Our previous studies in the same research trajectory were (1) ontology, epistemology, and discursiveness in teaching fundamental scientific topics, (Plakitsi & Kokkotas, 2010); (2) reflective, informal, and nonlinear aspects of argumentation in school practice (Plakitsi & Kokkotas, 2007), (3) enhancing teacher education through interpretive-philosophical mediation about the nature of science: The MAP prOject (Plakitsi & Kokkotas, 2006), and (4) discourse analysis (Piliouras, Plakitsi, & Kokkotas, 2007).

We also organize biennial national conferences in science for early childhood as well as biennial winter sessions for PhD candidates (Figures 11 and 12).

The former and the latter studies and academic activities show that we try to organize modern aspects of science education in a fruitful theoretical context that could push the theoretical and practical research in science education forward. This valid theoretical context with the dynamic characteristics of interactive systems of activities could be the CHAT context.

Overall, CHAT seems to be a coherent theoretical framework which can achieve the scope of real scientific literacy, enhance interdisciplinarity in Europe, and develop a new mentality that could reform science education from within.

The ATFISE PROJECT belongs to the third generation of activity theory, which is concerned with understanding and modeling networks of activity systems. The theoretical and methodological framework of analysis is the developmental work approach of Yrjö Engeström (1987, 2005). People participate in multiple activity systems within their local and global contexts, including online. International collaboration is an activity system that is also embedded within broader institutional, historical, and geopolitical contexts. A person engaged in one activity system is simultaneously influenced by other activity systems in which she or he participates. These influences both horizontal, happening across communities, and vertical, as social actions are also embedded within history, culture, and inequitable power relations that both influence the meaning, production, and shape of human activities. Within an activity system, all elements constantly interact with one another. Changes in the design of a tool may influence a subject's orientation toward an object, which in turn may influence the cultural practices of the community. Engeström (1987) called an activity system "a virtual disturbance-and-innovation-producing machine."

Figure 11. Biennial conference on science in and for early childhood with international participation (http://users.uoi.gr/5conns, webpage editor's archives).[5]

Figure 12. Biennial winter sessions for phd candidates in science education. (http://www.edife.gr, webpage editor's archives).

NOTES

[1] http://www.pisa.oecd.org/pages/0,3417,en_32252351_32235731_1_1_1_1_1,00.html
[2] See, for example, Jorde and Dillon (in preparation).
[3] See Van Eijck and Roth (2007).
[4] See Chapter 6 in this volume.
[5] + Poster design Nikos Giotitsas, biologist and PhD student.

REFERENCES

Engeström, Y. (1987). *Learning by expanding: An activity-theoretical approach to developmental research.* Helsinki: Orienta-Konsultit, Oy.

Engeström, Y. (1993). Developmental studies of work as a testbench of activity theory: The case of primary care medical practice. In S. Chaiklin & J. Lave (Eds.), *Understanding practice: Perspectives on activity and context* (pp. 64–103). Cambridge, MA: Cambridge University Press.

Engeström, Y. (1999). Innovative learning in work teams: Analyzing cycles of knowledge creation in practice. In Y. Engeström, R. Miettinen, & R. Punamaki (Eds.), *Perspectives on activity theory.* New York: Cambridge University Press.

Engeström, Y. (2005). *Developmental work research: Expanding activity theory in practice.* Berlin: Lehmanns Media.

Engeström, Y., Miettinen, R., & Punamaki R. (Eds.). (1999). *Perspectives on activity theory.* New York: Cambridge University Press.

Fleer, M., & Hedegaard, M. (2010). Children's development as participation in everyday practices across different institutions. *Mind, Culture, and Activity, 17*(2), 149–168.

Hedegaard, M. (2009). Children's development from a cultural-historical approach: Children's activity in everyday local settings as foundation for their development. *Mind, Culture, and Activity, 16*, 64–82.

Jorde, D., & Dillon, J. (Eds.). (in preparation). *A handbook of science education in Europe.* Rotterdam: Sense.

Kaptelinin, V., & Nardi, B. (2006). *Acting with technology: Activity theory and interaction design.* Cambridge: MIT Press.

Kaptelinin, V., Nardi, B. A., & Macaulay, C., (1999). The activity checklist: A tool for representing the "Space" of context, ACM /Interactions. *Methods & Tools, 6*, 27–39.

Lantolf, J. (2006). Sociocultural theory and L2: State of the art. *Studies in Second Language Acquisition, 28*(1), 67–109.

Plakitsi, K., & Kokkotas, V. (2010). Time for education: Ontology, epistemology and discursiveness in teaching fundamental scientific topics. *AIP Conference Proceedings, 1203*, 1347–1353. This paper was based on a presentation to the 1st International Conference of International Society for Cultural and Activity Research (I.S.C.A.R.), Seville, Spain, 2005.

Kravtsova, E. E. (2006). The concept of age-specific new psychological formations in contemporary developmental psychology. *Journal of Russian and East European Psychology, 44*(6), 6–18.

Kuutti, K. (1996). Activity theory as a potential framework for human-computer interaction research. In B. Nardi (Ed.), *Context and consciousness.* London: MIT Press.

Mwanza, D., & Engeström, Y. (2003). Pedagogical adeptness in the design of e-learning environments: Experiences from the Lab@Future project. In *Proceedings of e-learn 2003: International conference on e-learning in corporate, government, healthcare, and higher education* (Vol. 2, pp. 1344–1347). Phoenix, AZ.

Nardi, B. A. (1996). Activity theory and human-computer interaction. In B. A. Nardi (Ed.), *Context and consciousness: Activity theory and human-computer interaction* (pp. 69–103). Cambridge and London: MIT Press.

OECD. (2006). PISA 2006 results. Science Competencies for Tomorrow's World (2 Vols.).

Piliouras, P., Plakitsi, K., & Kokkotas, P. (2007). Sofia doesn't speak during team work. Using discourse analysis as a tool for the transformation of peer group interactions in an elementary multicultural science classroom. Paper presented to 12th Biennial Conference for Research on Learning and Instruction, EARLI 2007, Budapest, Hungary.

Plakitsi, K., & Kokkotas, V. (2006). Enhancing teachers' education through interpretive-philosophical meditation about the nature of science: The MaPrOject. Paper presented to the Joint North American – European and South American (N.A.E.S.A.) Symposium *Science and Technology Literacy in the 21st Century*, May 31–June 4, 2006, University of Cyprus. Proceedings, Vol. 1, pp. 200–211.

Plakitsi, K., & Kokkotas, V. (2007). Reflective, informal and non-linear aspects of argumentation in school practice. Yearbook of School of Education, University of Ioannina, B: 199-213.

Rogoff, B. (2003). *The cultural nature of human development.* Oxford, UK: Oxford University Press.

Roth, W.-M., & Lee, S. (2004). Science education as/for participation in the community. *Science Education, 88*(2), 263–291.

Roth, W. M., & Tobin, K. (2004). Cogenerative dialoguing and metaloguing: Reflexivity of processes and genres [35 paragraphs]. *Forum Qualitative Sozialforschung/Forum: Qualitative Social Research, 5*(3), Art. 7. Retrieved from http://nbn-resolving.de/urn:nbn:de:0114-fqs040370.

Roth, W.-M., & Tobin, K. (Eds.). (2005). *Teaching to-gether, learning together.* New York: Peter Lang.

van Eijck, M., & Roth, W.-M. (2007a). Keeping the local local: Recalibrating the status of science and Traditional Ecological Knowledge (TEK) in education. *Science Education, 91,* 926–947.

van Eijck, M., & Roth, W. M. (2007b). Rethinking the role of information technology-based research tools in students' development of scientific literacy. *Journal of Science Education and Technology, 16*(3), 225–238.

Vygotsky, L. S. (1998). *The collected works of L. S. Vygotsky* (M. J. Hall, Trans., Vol. 5, Child Psychology). New York: Kluwer Academic/Plenum.

Katerina Plakitsi
School of Education
University of Ioannina
Greece
kplakits@cc.uoi.gr

KATERINA PLAKITSI

2. CULTURAL-HISTORICAL ACTIVITY THEORY (CHAT) FRAMEWORK AND SCIENCE EDUCATION IN THE POSITIVISTIC TRADITION

Towards a New Methodology?

τὸ ἀντίξουν συμφέρον καὶ ἐ κ
τῶν διαφερόντων καλλίστην
ἁρμονίαν (καὶ πάντα κατ' ἔριν
γίνεσθαι).
Ηράκλειτος

Opposition brings concord.
Out of discord comes the fairest harmony.
Heraclitus[1]

INTRODUCTION

Two prominent methodological traditions, the positivistic and CHAT, developed in parallel and affected two approaches or paradigms to teaching and research (Kuhn, 1962). In this chapter we go through some basic elements of the two major traditions and we try to argue for a new methodology in science education, especially for early learners.

Recently, Kincheloe and Tobin (2009) published an article entitled "The much exaggerated death of positivism." From this rigorous study we selected some quotes in order to make our argument. It is true that in the twentieth century both psychology and education have been dominated by behaviorism, which is a form of empiricism as well as logical positivism. So, methodologically, and ….

from a behaviorist perspective, psychology is an objective experimental branch of natural science with a theoretical goal of predicting and controlling behavior. There is almost a preoccupation with method as a means of replica- ting results, and thereby identifying reproducible outcomes. The sources of behavior are external, belonging to the environment. A defining characteristic of behaviorism is a rejection of introspection and consciousness. If mental terms or concepts are used, they are to be translated into behavioral concepts. Causal regularities, laws and functional relations that govern the formation of associations are identified through experimentation in order to predict how

K. Plakitsi (ed.), Activity Theory in Formal and Informal Science Education, 17–26.

behavior changes and the environment changes. (Kincheloe & Tobin, 2009, p. 516)

The epistemology of positivism identifies valuable knowledge as that which involves objective information that reflects the real world. Kincheloe (2009) has classified six epistemological (with ontological dimensions) positivistic assumptions. These are Formal, Intractable, Decontextualized, Universalistic, Reductionistic, and One Dimensional. In Table 1, we assign these assumptions one by one to some assumptions which can be used with a CHAT approach to the topic of time. These are also informed from Plakitsi's doctoral thesis, "The Child's Conception of Time and Its Implications in Understanding Fundamental Scientific Concepts. An Inter-disciplinary Teaching Approach." The thesis belongs to a Piagetian perspective enriched by some multiplicity elements. In this section we attempt to show the potential for moving from the positivistic to CHAT paradigm. This development emerged because of the inappropriateness and inadequacy of the positivistic paradigm with respect to the whole pedagogical context in which science education occurs. Adapting CHAT to science education research, Table 1 provides a comparison of positivistic epistemology versus CHAT assumptions.

Table 1. Positivistic versus Cultural Historical Activity Theory epistemology.

Positivistic epistemology	*CHAT epistemology*
Formal We do research on the child's pheno-menology of the world, the object, time, etc. We use standardized tests, reliable questionnaires produced by a particular standard methodology, which is similar in any scientific research. We follow a rigorous step-by-step analysis. We teach the reliable scientific methodology to students as a step-by-step standard procedure.	**Nonformal** We do research with many different methods, without neglecting different research forms and traditions, for example, some qualitative methods. Researching and teaching are planned and replanned according to local societal conditions and the current circumstances.
Intractable For example, in the case of a child's conception of time, we investigate only the conventional aspect of time, which reflects the Newtonian concept of one unique and uniform time in the universe. All other aspects of time are excluded from the research planning. Even if we know about relativistic time, we teach and/or do educational research on conventional time, as it is the unique aspect of time. Also, conventional time is considered as stable –	**Intractable** Scientific concepts and also childhood are considered as ongoing processes; the teaching/researching refers to some milestones of their evolution. Changes considered as the target point of the teaching/researching.

an entity existing outside the child, in the external world (e.g., clocks). Furthermore, we tend to teach relativistic time only in high school and at the university level, while we have developed students with the Newtonian aspect of time being unique and independent of scientific evolution. Additionally, we partially teach time as a psychological infralogical structure (Piaget, 1972). Finally, any ongoing temporal event is outside of typical teaching/ researching.

Decontextualized

We do research on children's conceptions of time independently from children's sociocultural, economic, family, and school environments. When we teach conventional time using clocks, usually we focus on the mechanical clock; children have to learn hours, minutes, and seconds the same way. Rarely, we consider that some children from rural regions may already know the passage of time and the main daytime hours through sundials, sun movements, etc. And it is also rare to begin with the history of clocks, e.g., from water clocks in Egypt, candle clocks, clepsydras, and the pendulum to the modern mechanical clocks and then to quartz clocks and so on. Furthermore, the very interesting story about pendulums (Matthews, 2000) can show the children that major political concepts (science only for the elite and not for the masses) and religious beliefs (the creator as a clock maker) have influenced the standard measures and weights we have now.

Contextualized

We do research on chilrden's conceptions of time based on children's sociocultural, economic, family, and school environments. When we teach conventional time using clocks, children learn about time using their own pathways. We constantly take into consideration that some children from rural regions may already know a lot of temporal dimensions through rural activities. We recommend beginning with the history of clocks, e.g., from water clocks in Egypt, candle clocks, clepsydras, and the pendulum to the modern mechanical clocks and then to quartz clocks and so on. Furthermore, the very interesting story about pendulums (Matthews, 2000) can show the children that major political (science only for the elite and not for the masses) and religious concepts (the creator as a clock maker) have influenced standard measures and weights. Teaching/ researching is contextualized. Knowledge is always in context. Context analysis is important. Different contexts enrich teaching/ researching, interest, and challenge for change, while in the positivistic tradition different contexts cause research bias and teaching noise. Development is viewed as the ability of context transformation, and/or the movement from context to context, from system to system. There is teaching time, researching time, religious time, astronomical time, etc.

19

Universalistic

The basic principle here is that teaching leads to one form of knowledge: a true and stable knowledge. Science is being taught as the discovery of true knowledge, which exists in the real world. This knowledge is external and outside of the children. So the universe, the environment acts upon the children, and they internalize the impacts of the environment. This decontextualization was a prominent perspective in education, but we now know that scientific "truths" – as one example, the field of physics – have evolved considerably.

Multicultural

The central idea of this section is that there are many types of science, for example, western science, indigenous science, and personal science. Different ways of interpreting data lead to multiple world views that create unity from the differences. Time is one of example that shows how scientific paradigms (Kuhn, 1962) changed from one era to another. For example, Newtonian time is unique and uniform in the universe; the Kantian critique of pure reason supported this kind of scientific thought. Furthermore, no one teacher teaches anything about Bergson and his psychological aspect of time. The concept of time in Einstein's theory of relativity depends on the observer. Moreover, the conceptualization of time is totally different in quantum mechanics, where fundamental concepts such as "before" and "after" obtain a totally new meaning and where causality, as in classic physics, does not exist. Fortunately, the voices of great thinkers such as Popper and Kuhn reshaped modern scientific enterprise. Prominent psychologists such as Piaget also worked on the notion of time. In fact, we thus have various concepts of time: psychological, astronomical, and relative. Time has different meanings for Christians, Jews, Muslims, Buddhists, etc. Ultimately we have to accept that all kinds of time have scientific and cultural value.

Reductionistic

This means that only one research method is the scientific one, and any researcher can repeat the same research results any place in the world by following the same method.

During learning and teaching processes, we consider the achievement of the same objective as the one way to successfully orient an activity.

The researcher's origins, studies, way of thinking and observing, and the

Local

There is neither reductionist research nor one method of teaching because of the different forms of artifacts that mediate human activities. By keeping the local local, we can acquire a rich list of criteria, as well as ways of knowing and learning. We theorize that changes in systems of activities are successful and meaningful processes that occur during the activities. Change means

communities involved are from a positivistic research perspective. Objectivity and neutrality disguise the ethnocentric and colonial bias elements of research.	advancement. Succession of events means passage of time, and different ways of conceptualizing and measuring durations are approved.
Unidimensional Reality is unidimensional, so research on a child's conception of time therefore assumes the Western concept of time. Historically significant perceptions of time, such as those of the ancient Greeks (Parmenidis, Helakleitos, Plato, and Aristotle [Physics IV]) are excluded. Different meanings of time also exist among people of India, American Indians, and indigenous Australians. One voice is privileged. Multiple voices are opposed to the single correct scientific perspective. Even great modern researchers, who find language to communicate with general audiences about time according to quantum mechanics theory, adopt this perspective. Multiple voices are seen to threaten national security rather than a basis for reconceptualising national security as a unity of the different aspects.	**Multidimensional** Reality and environment are multi-dimensional and complex. We need new methodologies for teaching and researching in these interactive and progressive systems of relationships.

All epistemologies are explicit or implicit in any research planning or program. Teachers/researchers are often not conscious of their research bias. In this regard, Kincheloe and Tobin (2009) speak about "crypto-positivism."

> A central part of this crypto-positivism is adherence to a scientific method derived from the natural sciences and deemed necessary for a rigorous social science. (p. 514)

Positivism as logic for inquiry, and as a teaching background, can be considered as a part of a wider current of thought that reflects many colonial characteristics. This major tradition established the superiority of Eurocentrism, and is used to devalue other cultures using criteria of some encrusted prejudices. It is not fair to judge one culture by criteria from another. This practice promotes social exclusion while, as one example, the European Union made social inclusion a priority. Furthermore, some African, Indian, and Aboriginal cultures offer great benefits to western cultures, if the latter can make an effort to become familiar with indigenous knowledge.

Natural scientists are familiar with Einstein's theory about different observers in different points of the universe and with the absolutely revolutionary ideas of modern quantum mechanics theory. Simultaneously, natural sciences gave their reliable and superior method to social sciences. Consequently, we wonder why we still remain loyal to positivist thinking. There can be many answers to this question, derived from religion, politics, and economics. All these domains influence scientists and researchers while they do their work. All these epistemological anatreptic voices have been recorded by great thinkers such as Popper (1969), in *Conjectures and Refutations,* and Lakatos (1978), in *The Methodology of Scientific Research Programmes,* as well as Feyerabend (1975), in his work *Against Method.* In the concluding chapter, the latter wrote:

> To sum up: there is no 'scientific world view' just as there is no uniform enterprise 'science' .... Still, there are many things we can learn from the sciences. But we can also learn from the humanities, from religion and from the remnants of ancient traditions that survived the onslaught of Western Civilization. No area is unified and perfect .... [p. 249, emphasis added]

Pluralistic epistemologies support that knowledge is a constant process of creation rather than the static phenomenon of positivistic thinking, instead of the positivistic treatment of knowledge as a static phenomenon. The epistemologies, ontologies, and multiple research methodologies we embrace understand that educational phenomena are situated in environments constructed by their temporal interactions with the other dynamics in the world (Tobin & Kincheloe, 2006).

Overall, in this complex situation where the teachers/researchers must follow multiple voices and traditions and where researchers must reject old research and tools of observation in order to change themselves, we need to begin a dialogue within all nations to develop multiscience education systems. Aikenhead and Ogawa's clear description of the multiscience perspective (2007) is remarkable for its use of the words *science, knowledge, perceiving,* and *rational.*

> A rational perceiving of reality has three aspects: a process, a product (i.e., knowledge or action), and a cultural context defined by the people engaged in the perceiving. Ogawa (1995) considered three sciences: Eurocentric science, indigenous science, and personal science (a rational perceiving of reality unique to each individual, not discussed). (p. 544, brackets deleted)

Recapitulating, we need a different epistemology which will enable a shift from universalism to multiculturalism. Science curricula must include some traditional or indigenous forms of knowledge. Traditional ecological knowledge and typical science have been studied rigorously by van Eijck and Roth (2007). Concepts such as locality, multiculturalism, relativism, material reality, cultural reality, and plausibility have been deeply interpreted from a multicultural perspective. The authors prove the nonfeasibility of reductionism. Any type of nonwestern science fails to satisfy the criteria of good science because of the different frames of reference and the different relations with the material or the cultural reality (van Eijck & Roth, 2007, p. 931). Truth, knowledge, socialization, and enculturation are also discussed in this article. In fact, when structuring a curriculum, when making a project in schools, we must make choices—choices about the types of knowledge we shall use, rates of transition, and evaluating methods. All these are necessary epistemological options from a multiculturalism or universalism perspective. The authors argue for the heterogeneous character of knowledge as in the following quote:

> Knowledge, as integral to human being, is also a mêlée of voices, texts, procedures, tools, constructs, and so on; it exists only in and through its continuous production and reproduction in the concrete praxis of real human beings[2]. Even the most transcendental and deductive sciences, such as geometry, only exist in the dialectical relationship with human practices. As a continuous ongoing process subject to collective human practice, knowledge emerges and disappears as it is constructed and deconstructed, shaped and reshaped, produced and reproduced, forgotten and reminded, reinvented and taught. More so, even if we perceive knowledge as a body, as a singular identity in itself, it is so in the midst of other bodies of knowledge and therefore never on its own. In this sense knowledge is, like human bodies, singular plural in nature.[3] Consequently, knowledge also is essentially heterogeneous rather than homogeneous (van Eijck & Roth, 2007, pp. 932–933).

Accordingly, we need a different epistemology which treats reality both physically and culturally and incorporates the different types of knowledge as well as human practices by giving them the same value without superiorities and/or inferiorities. We propose that the framework provided by activity theorists is a coherent theoretical framework which establishes science education as participation in the community (Roth & Lee, 2004). Moreover, research on a nondualistic basis, as an inseparable aspect of a learning and doing system, is much more than a challenge. We are going to adapt activity theory as a designing tool for formal (schools) and informal (museums) science-education-places by using e-settings. This concept will advance the osmosis of a common European science learning culture. Modern schools and science museums in Europe organize many indoor and outdoor scientific activities based especially on e-learning technologies. In spite of all these efforts, there is no common European science learning environment informed by CHAT, especially for young learners (5–9 years old).

CULTURAL-HISTORICAL ACTIVITY THEORY

A New Paradigm

CHAT originated with Soviet thinkers who saw behaviorism and analytical psychology as unable to manage the material and cultural reality which was then on the scene. The concept of activity became very important in the societal setting, and the focus was on activity as the unit of analysis. Two kinds of activity element were distinguished: the cultural-historical and the material.

They studied activity mainly on the macrolevel, e.g., hunting, farming, constructing, building, producing. Human needs were at the top of the pyramid of social structures in a socialistic society. Accordingly, knowledge was a product, an outcome to be spread in the community, in which members/subjects act by dividing the work/labor and by following community rules and always using artifacts/tools. Tools might be mental, hands on, or simply a tool or machine.

The main role of the artifacts/tools is to mediate between subject and object. This mediation leads to outcomes that change society, the environment, and humans themselves. Thus, these social and environmental changes have reciprocal effects; social and environmental inputs require human accommodations and, in turn, human activities significantly influence the society and environment.

Knowledge also follows this schema of relations and is considered as part of object-oriented and artifact-mediated activity (Vygotsky, 1978). This schema imposed significant changes in scientific and teaching methodology. The unit of analysis is the activity, and the unit of learning community is the group. The focus was transferred from the individual to the system. The activity system became the central interactive unit with inputs and outputs. In this epistemology, one can find many similarities to the evolution and|or revolution of natural sciences. Dialectical relations form a scientific system as well as social activity systems. Dialectical relations between subject and object define activity (Roth & Lee, 2004; van Eijck & Roth, 2007).

Piagetian epistemology places individuals and their mechanisms of reasoning at its center. The prominent Piagetian method of clinical interview investigates the logical functions of individuals while excluding society from the interview settings. The teacher/researcher seems to examine an individual's reasoning or concept-ualization a posteriori, that is to say, at the end of an individual's activity with the society or the environment.

Instead of this approach, CHAT inspired teachers/researchers to study human learning as being human (biological, evolutional), belonging (societal and environ-mental), and becoming (societal and environmental) (Lee and Roth, 2003). Without a priori axioms a new epistemology emerged: an epistemology of Being and Time, as Heidegger proposed (Heidegger, 1982, 1992). In CHAT, knowledge is always in context. This context gives meaning to the artifacts.

With the passage of time, artifacts become better and better, while human experience with their use in turn reshapes artifacts. Then, the elaborated artifacts modify or totally change the activity in an eternally ongoing process. History and culture form scientific activities, which in turn change history and culture and vice versa in a dialectical way of being and uniting the opposites, as articulated by Heraclitus in his doctrine of change: Opposition brings concord. Out of discord comes the fairest harmony (Heraclitus, fragment 98).

Roth and Lee (2007) argue that CHAT allows us to approach and analyze typical activity systems (praxes, contexts). The activity systems can be, for example, science, environmentalism, and indigenous knowledge systems.

At the macrolevel, when moving from actions to activity, we have expansive learning (Engeström & Sannino, 2010). The theory of expansive learning focuses on learning processes in which the very subject of learning is transformed from isolated individuals to collectives and networks (p. 5). Expansive learning is manifested primarily as changes in the object of the collective activity (p. 8).

Boundary crossing involves collective concept formation (p. 12). Expansive learning takes place because historically evolving contradictions in activity systems lead to disturbances, conflicts, and double binds that trigger new kinds of actions among the actors (p. 18).

In conclusion, we argue that it is obvious that we need a new methodology in/for teaching and researching. Maybe we can start with concepts such as "communities of practice," "collaborative planning," "learning communities," and "socioemotional learning," which are enabled in many modern school curricula. We have also to work on new values. Whether the current public schools continue to have a monolithic and deterministic bias or whether new dialogues will emerge and create novel institutional structures depends upon our efforts. A current article by Jahreie and Ottesen (2010) discusses "third spaces" and "learning spheres." These articles, and the very important schema of Marianne Hedegaard (2009) (see Chapter 1), move the discussion beyond Engeström's triangle of activity. We can do educational research in different local settings and keep the local local. Also we can study children acting through participation in different institutions and contexts. Finally, we can focus learning spheres in order to study knowledge acquisition. We must change ourselves in order to study changes in social, environmental, and personal systems of activities.

NOTES

[1] Fragment 98, as translated by Philip Wheelwright, in Wheelwright, P. (1966). *The Presocratics.* Indianapolis: ITT.

[2] See Chapter 7 entitled: In praise of the mêlée. Toward a new conception of scientific identity and literacy. The chapter is published in the book: Roth, W.M., van Eijck, M., Reis, G., & Hsu, P-L. (2008). *Authentic Science Revisited.* Rotterdam: Sense.

[3] Roth, W-M. (2006). *Learning science: A singular plural perspective.* Rotterdam, The Netherlands: Sense Publishers.

REFERENCES

Aikenhead, G. S, & Ogawa, M. (2007). Indigenous knowledge and science revisited. *Cultural Studies of Science Education, 2,* 539–620.

Engeström, Y., & Sannino, A. (2010). Studies of expansive learning: Foundations, findings and future challenges. *Educational Research Review, 5,* 1–24.

Feyerabend, P. (1975/1993). *Against method.* London: Verso.

Hedegaard, M. (2009). Children's development from a cultural-historical approach: Children's activity in everyday local settings as foundation for their development. *Mind, Culture, and Activity, 16,* 64–82.

Heidegger, M. (1982). *The basic problems of phenomenology* (A. Hofstadter, Trans.). Bloomington, IN (German original first published in: Martin Heidegger, Gesamtausgabe, II. Abteilung: Vorlesungen 1923–1944, Bd. 24, Frankfurt a.M. 1975).

Heidegger, M. (1992). *Being and time* (J. Macquarrie & E. Robinson, Trans.). London: SCM Press, 1962; re-translated by Joan Stambaugh (Albany, NY: State University of New York Press, 1996).

Jahreie, C. F., & Ottesen, E. (2010). Construction of boundaries in teacher education: Analyzing student teachers' accounts. *Mind, Culture, and Activity, 17*(3), 212–234.

Kant, I. (1787/1933). *Critique of pure reason* (2nd ed., N. K. Smith, Trans.). London: Macmillan (first published: London 1929; German original first published: Riga 1781[A]; 1787[B]).

Kincheloe, J. L., & Tobin, K. (2009). The much exaggerated death of positivism. *Cultural Studies of Science Education, 4*, 513–528.

Kuhn, T. (1962/1996). *The structure of scientific revolutions.* US: University of Chicago Press. Translation in Greek: Kuhn, T. (1981). *Η δομή των επιστημονικών επαναστάσεων.* Θεσσαλονίκη.

Lakatos (1978). *The methodology of scientific research programmes: Philosophical papers* (Vol. 1, J. Worrall & G. Currie, Eds.). Cambridge: Cambridge University Press. Translation in Greek: Lakatos, I. (1986). *Μεθοδολογία των Προγραμμάτων Επιστημονικής Έρευνας.* Μτφρ. Αιμ. Μεταξόπουλος, Σύγχρονα θέματα, Θεσσαλονίκη.

Lee, S. H., & Roth, W.-M. (2003). Becoming and be-longing: Learning qualitative research through legiti-mate peripheral participation. *Forum Qualitative Sozialforschung/Forum: Qualitative Social Research, 4*(2).

Matthews, R. M. (2000). *Time for science education: How teaching the history and philosophy of pendulum motion can contribute to science literacy.* New York: Kluwer Academic/Plenum Publishers.

Piaget, J. (1969). *The child's conception of time.* Translation from French. (1927). *Le development de la notion de temps chez l'enfant.* PUF, London: Pomerans, A. J. Routledge & Kegan Paul.

Piaget, J. (1972). *Psychology and epistemology: Towards a theory of knowledge.* London: Penguin University Books.

Popper, K. (1969). *Conjectures and refutations.* London: Routledge, Kegan Paul.

Roth, W. M., & Lee, S. (2004). Science education as/for participation in the community. *Science Education, 88*, 263–291.

Roth, W.-M., & Lee, Y. J. (2007). "Vygotsky's neglected legacy:" Cultural-historical activity theory. *Review of Educational Research, 77*, 186–232.

Tobin, K., & Kincheloe, J. (Eds.). (2006). *Doing educational research: A handbook.* Rotterdam: Sense Publishers.

van Eijck, M., & Roth, W.-M. (2007). Keeping the local local: Recalibrating the status of science and Traditional Ecological Knowledge (TEK) in education. *Science Education, 91*, 926–947.

Vygotsky, L. S. (1978). *Mind in society: The development of higher psychological processes.* Cambridge, MA: Harvard University Press. Edited by M. Cole, V. John-Steiner, S. Scribner, E. Souberman. Published originally in Russian in 1930.

Wheelwright, P. (1966). *The Presocratics.* Indianapolis, IN: ITT.

Katerina Plakitsi
School of Education
University of Ioannina
Greece
kplakits@cc.uoi.gr

KATERINA PLAKITSI

3. TEACHING SCIENCE IN SCIENCE MUSEUMS AND SCIENCE CENTERS

Towards a New Pedagogy?

INTRODUCTION

This paper is a review of ten years of research on the relationship of science educa-tion and museum education in the context of learning sciences and lifelong learning. A multiset of papers – more than 25 – from different institutions, such as schools, museums, and universities, is reviewed through the prism of Cultural Historical Activity Theory (CHAT).

FORMAL AND INFORMAL SCIENCE EDUCATION

At the beginning of this chapter we clarify the use of the terms *formal* and *informal science education.* We call formal science education any typical learning environ-ments, approaches, or contexts that are limited to the scope and content knowledge imposed by the mainstream educational system, usually the national curriculum. They refer both to compulsory and postcompulsory education. On the other hand, informal science education consists of free-choice learning environments, approaches, or contexts that can be connected or not to the national curriculum. They are systematic and cumulative aspects of everyday experiential learning. They are lifelong learning environments and, as they are strongly embedded in everyday experience, they play an important role in human learning. Between the two extremes, formal and informal, there is a continuum consisting of many medium contexts, more or less organized and related to the typical national educational system. They are also called nonformal settings. In our study we shall use the term *informal* to describe any nonformal setting, approach, or context of science education. The main issue is that formal and informal science education are en face views of the same face.

Usually, a combination of formal and informal science education is represented by the term "science in society," which is on the frontier of the global educational scene. So, a dialectical relationship between science and|for society became a classic, lifelong, and worldwide tradition. Science in Society mainly means learning science in science museums and science centers, where visitors from the general public are also involved. Furthermore, many schools do their science courses in science museums and science centers. There are many indoor activities where students and teachers interact with each other and learn science, as well as

K. Plakitsi (ed.), Activity Theory in Formal and Informal Science Education, 27–56.

many outdoor scientific activities where teachers, students, and parents interact during their daily experiences as citizens.

In recent years, in Europe, science in society has become a priority through Frame Program Seven.[1] Research in this domain aims for a more dynamic governance of the relationship between science and society. It includes ethics, culture, informal debates, strengthening the role of women in science, support of formal and informal science education in schools as well as through science centers and museums and other relevant means, and, focuses on science and society communication.

Many prominent scholars, including Roth and McGinn (1997), proposed deinstitutionalizing school science education. This proposal was based upon the theory of situated cognition. The research trajectories of Wolf-Michael Roth and Ken Tobin have had greatly influenced this effort. Both scholars posed the necessity of expanding the concept of science education to include cultural acquisition and participation in the community (e.g., Roth & Tobin, 2002; Roth, 2010). Furthermore, teaching science in science museums and science centers is strongly connected with the sociocultural aspects of science education. We support applying the CHAT framework in science education so we can expand the borders of our pedagogical knowledge. A new pedagogy can emerge from the combination of the two relevant paradigms: science education and museum education. In the CHAT framework, tools from the history and philosophy of science as well as epistemological and pedagogical rules, limits, and principles mediate the subject–object interactions and create culture, which in turn becomes structure. This procedure in science museums and science centers can be more liberating and more motivating and lacks any form of symbolic violence (Bourdie's terminology).[2] This becomes very important when students cross the borders by horizontal or vertical movements among different interactive systems of various subcultures (subculture of science, personal subculture, societal subculture, etc.; see Aikenhead, 2007). Furthermore, learning in science museums and science centers can physically and logically be embedded in the CHAT context, since museum exhibitions are strong cultural tools and play a central mediative role in learning and culture making.

DIALOGICAL TURN IN SCIENCE EDUCATION AND NEW ROLES FOR SCIENCE MUSEUMS AND SCIENCE CENTERS

According to the International Council of Museums (ICOM),

A museum is a non-profit, permanent institution in the service of society and its development, open to the public, which acquires, conserves, researches, communicates and exhibits the tangible and intangible heritage of humanity and its environment for the purposes of education, study and enjoyment. (ICOM Statutes, adopted by the 22nd General Assembly, Vienna, Austria, 24 August 2007)

Museums represent a major public social investment that has been made by most countries in the last 200 years or more. Their influence on society is powerful, in either peace time or war. The destruction of many museums, as well as strenuous efforts for museum protection, has been recorded during recent wars. Precious artifacts have been moved, protected, or stolen during those periods. Usually, a museum's impact on the general public is underestimated. However, in the United States, the American Association of Museums claims that annual museum attendance is close to a billion visits a year.[3] The AAM estimates the number of "infinitely diverse" museums in the 15,000–16,000 range, and notes that the draw of museums is greater than the annual attendance at professional sports events. The term "museum" is also described by an even looser definition: "any institution, built, or interpreted environment that may have an educational role, whether education is part of its mission statement or not" (Rennie & Johnson, 2004, p. S4).

Consequently, some of the major categories of museums are museums of art, historical museums, and science museums but also museums of "general" content. Some independent exhibits, for example, the Statue of Liberty itself, can be considered by some as a "museum." According to another categorization (Koliopoulos, 2005), museums of natural sciences and technology can be classified in four categories:

1. The museum-institution, which expresses the traditional form of museum and which, as time passes, transforms itself from a static to a functional entity, which incorporates intense educational activities.
2. The virtual museum, which is a museum without walls where networking and new forms of communication dominate.
3. The children's museum, which primarily serves children, in particular very young children.
4. The local museum (or museum in situ), which is connected closely with the local natural and social environment.

Examples from this classification (Koliopoulos, 2005), are:

1.1. Museum-Institution: Collections

Figure 1. (a) London Science Museum, U.K.: Apollo 10 mode. (b) The Future of Biometrics in the new Antenna Gallery.
(Figures 1a, b are downloaded from the official site from London Science Museum: http://www.sciencemuseum.org.uk/Centenary/Home/Icons/ Apollo_10_Capsule.aspx, http://www.info4 security.com/ story_attachment.asp?storycode= 4114235&seq=2&type=P&c=2 (accessed 2011-8-30).

1.2. Museum-Institution: Experiments-Inventions

Figure 2. (a) Exploratorium San Francisco, U.S., Inventor. (b) Palais de la Découverte, Paris, France, electrostatique. (c) Nemo, the Netherlands, water force.
(Figures 2a, b, c are downloaded from the official site of Exploratorium San Francisco: http://www.exploratorium.edu/index.php, accessed 2011-8-30.)

1.3. Museum-Institution: Cultural Centers

Figures 3. (a) Entrance Hall of the cultural, scientific center, Cité des Sciences et de l'Industrie, Paris. (b) Eugenidis Foundation, Planetarium, Athens, Greece.
(Figure 3a is downloaded from the official site of Cité des Sciences et de l'Industrie http://www.cite-sciences.fr/fr/cite-des-sciences/; Figure 3b is downloaded from the official site of Greek Eugenidis Foundation http://www.eugenfound.edu.gr/ frontoffice/portal.asp?cpage=NODE&cnode=1, accessed 2011-8-30.)

2. The Virtual Museum

Levy (2001) distinguishes three eras that are shaped by different ways of knowledge acquisition: the era of oral communication, the era of writing, and the era of typography. We can say that the endless electronic library today constitutes a fourth era, in which data bases, simulations, electronic conferences ensure better knowledge of the world than did the theoretical abstractions of the past.

An extension of this digital culture is web communication, which dominates our everyday life. Another extension is the creation of many virtual museums of natural science and technology that serve either adults or children. The main contribution of virtual museums is the virtual representations and intense virtual experiences that they make available to their public. Systems of virtual reality and augmented reality are central in each modern museum of natural sciences and technology, as well as in many interactive parks.

According to Levy, the word virtual has three dimensions–one technological, one modern, and one philosophical–and together they provide the charming characteristics of any virtual museum. The technical systems of virtual reality play an important role for the public too. The mediative tools usually include the central processor, a three-dimensional mouse, glove, joystick or keyboard, and a system of optical presentations. The optical presentations may include simple pictures on a level screen, a stereoscopic picture with a sense of depth, and, finally, when the user wears a helmet or protective binoculars, two different pictures that are combined in the optical system with a resultant perception of depth and locomotion. It is important that the user's entire body is engaged as a pilot as he or she performs many interactions in virtual space.

Finally, many museums present their exhibits via avatars and guide robots. The robots take pictures from the exhibits that appear on the museum's web page and are thus available to visitors. The robot can also welcome the visitors, lead them to exhibits, and provide explanations.

Some useful links for virtual museums are:
- International Council of Museums, http://icom.museum/vlmp/ (accessed 27/7/2010)
- The Virtual Library of Museums in USA, http://museumca.org/usa/ (accessed 27/7/2010)
- European Network of Science Centres and Museums, www.ecsite.net, (accessed 27/7/2010)
- Association of Science –Technology Museums, U.K. www.astc.org (accessed 27/7/2010)
- Association of Children's museums, Hands on Europe, www.hands-on-europe.net, (accessed 27/7/2010)

Some examples of museums with web cameras and/or robots are:

- Deutsches Museum Bonn, Germany: http://www.deutsches-museum.de/en/bonn/information/ (accessed 14-12-2010).
- National Museum of Nature and Science, Tokyo, Japan: http://www.kahaku.go.jp/english (accessed 14-12-2010).
- Haus der Geschichte der Bundesrepublik Deutschland: http://www.hdg.de
- La città dei Bambini, Genova, Italy http://www.cittadeibambini.net
- Museum für Kommunikation Berlin, Deutschland
- www.museumsstiftung.de/berlin.html (accessed 14-12-2010).
- State Darwin Museum, Moscow, Russia http://www.darwin.museum.ru (accessed 14-12-2010).

31

- Swiss Science Center, Technorama, http://www.technorama.ch (accessed 14-12-2010).
- Virtual museum of Canada, http://www.museevirtuel-virtualmuseum.ca/index-eng.jsp (accessed 27/7/2010) (accessed 14-12-2010).
- Institute of Computer Science, Crete, Greece
- http://www.ics.forth.gr/cvrl/robots.html#lefkos (accessed 14-12-2010). Robot Lefkos guides visitors in many Greek museums as the National History Museum of Crete, Herakleion, Greece (http://www.ics.forth.gr/cvrl/PHOTOS/nhm.html (accessed 14-12-2010).
- Memorial University – Ocean Sciences Centre, http://www.mun.ca/osc/seal-lab/webcams_0.php (accessed 14-12-2010).
- The Robot Zoo, London
- http://www.evergreenexhibitions.com/exhibits/robot_zoo/index.asp (accessed 14-12-2010).
- California Academy of Science, USA
- http://www.calacademy.org/webcams/penguins/ (accessed 14-12-2010).
- Science Museum of London
- (http://www.sciencemuseum.org.uk/educators/classroom_and_homework_resources/resources/robot_bug.aspx (accessed 14-12-2010).

3. Children's Museums

The Association of Children's Museums in the United States and Canada (ACM), http://www.childrensmuseums.org/index.htm, (accessed 6-8-2010) reports that:
- There are more than 515 ACM members including children's museums, businesses, individuals, and museums with programs for children.
- According to 2007 data on ACM, more than 30 million children and families visited children's museums annually.
- Outreach programs in ACM member museums extended to nearly 4 million people in 2007.
- The largest children's museum is The Children's Museum of Indianapolis (Indiana), which has a total of 433,000 square feet.
- The oldest children's museum is the Brooklyn Children's Museum (New York), which opened in 1899; the children's museum field is 111 years young! An d the only LEED certified children's museum in XXX

ACM is recognized as a global leader, advocate, and resource among organizations serving the learning needs of children and families. Also, ACM's mission is to build the capacity of children's museums to serve as town squares for children and families where play inspires creativity and lifelong learning.

Boston Children Museum's history stresses the importance on the educational role of children's museums in natural and environmental sciences (http://www.bostonkids.org/about/media_timeline.html accessed 6-8-2010).

In 1962 the New Museum director Michael Spock (1962–1985) was the basic founder of the American Association of Youth Museums (now Association of

Children's Museums). At the same time, the "Do Not Touch" sign was forever removed, according to Michael Spock's desire to create a museum where children would learn by interacting with the exhibits themselves. The first interactive exhibit, 'What's Inside,' allowed visitors to look at the inside of a baseball, a Volkswagen, and more. In 1972, the Museum was accredited by the American Association of Museums. Meanwhile, a recycling program was developed which offered unusual art materials at low cost.

The first exhibit for children 5 and under, PlaySpace, opened in 1978 and allowed children and parents to learn together. Bubbles, was an interesting science exhibit that was designed to familiarize children with some simple principles of geometry and physics through play and experimentation. In 1986, the new director Kenneth Brecher focused on the Multicultural Initiative to ensure that the Museum was increasingly reflective of a racially, ethnically, culturally, and economically diverse audience. Lou Casagrande, Ph.D., became the Museum's newest president and CEO. Casagrande organized the Museum into three strategic centers – Visitor Center, Teacher Center, and Early Childhood Center – to respond to the demands of education reform and the critical needs of the urban community. Finally, in 2008 Boston Children's Museum was recognized as LEED® Gold certified by the U.S. Green Building Council. The environmental dimension of the museum is represented by the recycle project and green building certification. Science (natural sciences) is obvious in any museum project as, for example, in the bubbles exhibit, NSF collaboration, the What's Inside project, and Playspace, with children and parents interacting with exhibits themselves. All these hands-on activities can make science interesting and children enthusiastic about science. In parallel, children's museums participate in every educational reform, and they especially develop joint projects between schoolscommunity-museum.

This is a sociocultural function of a museum, and children's museums are especially involved in such collaborations. In addition, teachers and scientists can find an excellent authentic learning environment in which to develop their theories or practices. For us, as science educators and researchers, the major issue is the interpretative paradigm in this enterprise. Cultural-historical activity theory fits better in children's museums and can make them bold tools for achieving great societal outcomes (see also Chapter 1).

In children's museums a child can grow up with science as a way of life, moving towards peace and sustainability. And this potential becomes the core objective for the educational activities of a new children's museum.

Figure 4. (a) Brooklyn Children's Museum, founded in 1899. (b) Indianapolis Children's Museum, U.S.
(Figure 4a is downloaded is downloaded from the official site of the Brooklyn Children's Museum http://www.brooklynkids.org/index.php; Figure 4b is from the Children's Museum of Indianapolis http://www.childrensmuseum.org/, accessed 2011-8-30.)

Figure 5. a. Mirdif City Centre Children's Museum, Dubai, United Arab Emirate., b. Entrance of the Cité des Enfants, Paris, France.
(Figure 5a is from http://www.mirdiffcitycentre.info/mirdiff-city-centre-aquaplay.php; Figure 5b is from http://www.cite-sciences.fr/fr/cite-des-sciences/ contenu/c/1248104303612/cite-des-enfants/, accessed 2011-8-30.)

4. The Local Museums (Museums In-Situ)

Local museums are very important for local communities, and they can promote the teaching and learning of science as a means of participating in the community. An example is the collection of Mediterranean olive oil museums located in specific places as indicated in the following map.

In Greece, for example, the Museum of the Olive and Greek Olive Oil in Sparta highlights the culture and technology of the olive and olive production. The olive's needs, uses, and symbolism make it one of the most important agricultural products, with a distinguished role in the economy of each historical period. In addition, olives are part of Greek and Mediterranean identity. Unique in Greece, the museum is located in the heart of Laconia, one of the main olive-producing areas in Greece. The local community, local museum, local history, local science, etc., form a localized but authentic learning environment for science education.

Figure 6. Mediterranean olive oil museums.
(http://www.oliveoilmuseums.gr/echome.asp?lang=2, accessed 2011-8-30.)

Beyond the previous review we posit (Kokkotas & Plakitsi, 2005) the following:

Museum Education includes education:

– within the Museum, both old and modern museums, that aim to enculturate the general public,
– that uses the Museum as a means for multicultural and interdisciplinary learning,
– that places the culture itself in the framework of cultural relativism, where each culture has its own value that is equivalent to the value of any other culture, for sustainability.

Science Education includes:

– laboratory work, science in schools, field work in nature using the systematic processes of scientific method and aiming for scientific literacy for all citizens,
– uses science as a means, to develop educational effectiveness, in the framework of interdisciplinarity. In this approach we use scientific methods that traditionally use natural scientists for the study of subjects through the perspectives of different scientific fields, thus ensuring inter-, intra-, infra-, and cross-correlations,

35

– aims at education for citizenship, through achieving scientific and technological literacy for all. This means that each citizen will internalize certain basic cultural products from science and technology, which he or she will use in his or her social interactions, decision-making, and self-fulfillment.

Overall, a necessity for cultural and value relativism has emerged. The foregoing aspects of science education presuppose a relativistic approach to values, according to which each distinguishable scientific branch has the same value as any other. It follows, then, that the tools and the strategies used in science education are like those of museum education, such as role playing and debate. Common objectives as well as common applications can be found between science and museum education.

On the other hand, new roles for science education incorporate (1) new approaches to learning and teaching science (constructivism, sociocultural interactions), (2) interdisciplinarity, (3) student research projects, (4) inquiry teaching, (5) collective curriculum design, (6) research, (7) dance, theater, music, movement, and so on, (8) outreach activities: museums, community centers, etc., and (9) web communication.

New roles for museum education include: (1) the museum's shift from exhibit-oriented institutions to visitor-oriented cultural centers that promote meaning making, (2) dynamic networks of interdisciplinary educational programs, (3) educational materials and resources, (4) hands-on and (5) minds-on activities, (6) research, (7) dance, theater, music, movement, and so on, (8) outreach activities: classrooms, community centers, etc., and (9) websites.

A major innovation is that science museums and science centers are attempting to shift from being centers of collections of exhibits displayed in glass cases and static shows to being centers of interaction, activities and ideas that help visitors investigate and study not only physical phenomena but also modern socioscientific issues (Koster, 1999). At the same time they have advanced their educational role, which is no longer formal education in parallel with or as a basic component of formal education.

We can classify the evolution of science museums in three periods, from the early beginnings of the 20th century till today. During the first period, until the mid-1960s, science was considered a specific form of knowledge that could only be transmitted by experts from the cathedra to nonexperts. This was the period in which the miracles of science and the giants of science were admired. During the second period, that takes us up to the middle of the 1980s, scientific logic and methodology became privileged fields to be learned from nonexperts through their active participation in scientific activities. During the third period, science has been embedded in its social and cultural contexts. The three periods can be interpreted as the result of a decreased authority gap between experts and nonexperts (Dimopoulos, 2006).

In the constructivist and postconstructivist frameworks, we adopt the position of epistemological pluralism. In this way teachers can change their practices when teaching science, and this is important, since the teacher is the main mediator in student learning and citizenship.

As a complement to this perspective, some postmodern constructivist scholars (Taylor and Willison, 2002), have supported the following positions:
- The dialectical complementarity of epistemologies as the dynamic of integration, in contrast to Cartesian dualism.
- The metaphor, as a frame of reference, facilitates the crossing of borders and avoiding the obstacles of scholasticism that tend to diversity.

According to the above, by bridging humanitarian and science education, which is the basis of the argument for epistemological pluralism, environmental history can strengthen a society's valuation of its ecological heritage and make obvious the connection between human activity and environmental quality.

Conditions for "bridging" are:
- The extended definition of a "museum."
- Modern interdisciplinary instructive approaches and practices.
- The training of museum employees.

Koster (1999) suggests that an emerging set of seven attributes will largely define the next generation of science centers. Of these seven attributes, three speak directly about the role of topical and issues-based exhibitions. He predicts that science centers will focus on (pp. 291–292):

1. Missions centering on integrated interpretations of science-technology-society (STS) issues that focus more on today and tomorrow than on yesterday and that entertain multiple points of view.
2. Developing topical multidimensional experiences that will promote the perception of science centers as worthwhile lifelong learning resources.
3. Serving as a neutral ground for airing society's most vexing issues related to science and technology.

Koster (2006)[4] argued for the relevant museum – the relationship between relevance and sustainability – and suggested some indicators that would move museums towards relevancy. We choose three of them in order to illustrate the new societal role of modern museums.

1. Is your museum's mission statement explicit about the way(s) in which the institution aspires to be of tangible social and/or environmental value?

2. Does your museum periodically assess its mission in relation to changes in the external environment so as to identify better ways to direct its expertise and resources to areas of beneficial learning by the primary audience(s)?

3. Does your museum monitor and apply research findings in self-guided and mediated learning styles, and in allied and competitive fields such as formal education and other learning experiences? (list is quoted from http://www.aam-us.org/pubs/mn/ MN_MJ06_RelevantMuseum.cfm accessed 14-12-2010)

The modern context of the cultural–dialogical turn of science education leads science museums and science centers to become fruitful learning environments that promote meaningful learning in science. Pedretti (2004) describes the properties of issues-based exhibitions:

They present robust views of science, by promoting healthy public debate about complex socioscientific subject matter (Pedretti, 2004, p. S44).

These exhibitions promote "public awareness about science that entails understanding and critiquing the nature, processes, and achievements of science, as well as the interactions between science, technology, society and environment (STSE). (Pedretti, 2004, p. S44)

Also, these exhibitions:

... provide experiences beyond usual phenomenon-based exhibitions and carry the potential to enhance learning by personalizing subject matter, evoking emotion, stimulating dialogue and debate, and promoting reflexivity. Thus they serve as excellent environments in which to further explore the nature of learning in these settings. (Quoted from Pedretti, 2004, p. S45)

In a later work, Pedretti (2008) tests how issues-based installations promote science knowing and learning in an interdisciplinary STSE context. The authors focus on:

1. Teacher challenges and strategies in planning for and implementing an issues-based STSE approach;
2. Knowledge and skills gained by students in an STSE curriculum emphases; and
3. Pedagogical differences between an issues-based approach and other approaches used in science teaching (list is quoted from Pedretti, 2008, p. 945).

LIMITED INTERPRETATIVE PARADIGMS

Until now, interpretive theories for learning science in science museums and science centers have been limited to Falk and Dierking's (1992) and Hein's explanatory model (1998).

Falk and Dierking's 1992 Interactive Experience Model describes the museum experience in terms of interactions among physical, social, and interpersonal contexts. Falk and Dierking (2000) put forward their Contextual Model of Learning as "a device for organizing the complexities of learning within free-choice settings." The Contextual Model of Learning draws from constructivist, cognitive, as well as sociocultural theories of learning. Constructivist theories interpret the personal context, which in its turn represents the sum total of personal and genetic history that an individual carries with him/her into a learning situation. Consequently, prior knowledge and experience influence learning in museums. Social context is the core context of humans. According to many prominent socioculturalists, we are all products of our culture and social relationships. Hence, it is widely accepted that museum learning is socioculturally situated. Many factors affect this kind of learning, for example, the cultural value placed upon free-choice learning as well as the cultural context of the museum within society (Hooper-Greenhill, 1992). In fact, there are many interactive systems of activities among museum explainers, guides, demonstrators, performers, and visitors that lead to different outcomes in student learning. Finally, learning always occurs within the physical environment; thus the physical context of the museum itself is crucial for museum learning. We can see the physical context at the macro level as the large-scale properties of space, lighting, and climate as well as in the meso level, such as exhibitions and objects contained within, and at the micro level when analyzing the interactions among subjects, objects, rules, tools, and community. Falk and Dierking, 2000 claim that since museums are typically free-choice learning settings, the experience is generally voluntary, nonsequential, and highly reactive to what the setting affords.

Falk and Storksdieck (2005, p. 747) consider 12 key factors or, more accurately, suites of factors, that emerged as influences on museum learning experiences. These 12 factors are:

Personal Context
1. Visit motivation and expectations,
2. Prior knowledge,
3. Prior experiences,
4. Prior interests,
5. Choice and control,
Sociocultural Context
6. Within group social mediation,
7. Mediation by others outside the immediate social group,
Physical Context
8. Advance organizers,
9. Orientation to the physical space,
10. Architecture and large-scale environment,
11. Design and exposure to exhibits and programs,
12. Subsequent reinforcing events and experiences outside the museum

These factors differentiate the learning role for different visitors and venues, such as science centers, natural history museums, zoos, planetaria, and nature centers. Falk and Storksdieck (2005, p. 748) consider the following questions. "How do specific independent variables individually contribute to learning outcomes when not studied in isolation? Does the Contextual Model of Learning provide a useful framework for understanding learning from museums?"

They posit that:

Environments developed to support real-world learning such as museums, are not mere backdrops for supporting the transmission of knowledge, they are what Barab and Kirshner (2001) call "dynamical learning environments." As such, these settings are always multidimensional, dynamic, and complex (cf. Brown, 1992; Cobb et al., 2003; Collins, 1999). Thus the real take-away message of this article is that simple, reductionist, linear approaches to affecting and understanding learning from museums will simply not suffice. An awareness of this reality has begun to creep into school-based learning research as well, most notably under the banner of "design research" (cf. Brown, 1992; Cobb et al., 2003; Collins, 1999). Only by appreciating, and accounting for the true complexities of the museum experience will improved facilitation and understanding of learning from museums emerge. (Falk & Storksdieck, 2005, p. 772)

Hein (1998) proposed another model of museum learning. As every educational theory is built upon a knowledge theory and a learning theory, Hein distinguishes four types of museum learning: traditional lecture and text (leading to the Systematic Museum), Discovery Learning (the Discovery Museum), stimulus-response (the Orderly Museum), and constructivism (the Constructivist Museum.) He strongly argues that the most powerful and appropriate educational theory and practice for museums is constructivism (Hein 1998). We believe the most important contribution of constructivist theory to museum education is the center of the theory, which is meaning making. And this meaning making, based on cultural psychology, leads to hermeneutics and then to sociocultural theory. Bruner, in *Acts of Meaning* (1990) and in *The Culture of Education* (1996), highlights these issues. Especially, he describes the limits of the common rules of the information systems to cover the various forms of meaning making. The typical linear model of input information that after proseccing becomes comprehensible outputs cannot cover the messy, ambiguous, and context-sensitive processes of meaning making (p. 5). And the American psychologist highlights that:

When education narrows its scope of interpretive inquiry, it reduces a culture's power to adapt to change. (p. 15)

The matter of fact is that most psychologists have been educated in positivism, so they usually cannot adapt and adopt notions as belief, desire, and intention as explanations (Acts of Meaning, p. 15). According to our opinion, George Hein is strongly anchored to Dewey's perspective that any meaning making is established through experience. Also, he sees culture as a vehicle that any learner or visitor carries as mental baggage in a science center. But constructivism still keeps alive the true and false Kantian dualism.

Bruner, in his work The Culture of Education (1996, Ch. 2.4, p. 115, "Narratives of Science"), highlights the following topics:

What he knew was that science is not something that exists out there in nature, but that it is a tool in the mind of the knower – teacher and student alike ... There are lots of different ways of getting to that point and you don't really ever get there unless you do it, as a learner, on your own terms ... *For, in effect, a curriculum is like an animated conversation on a topic that can never be fully defined, although one can set limits upon it. I call it an 'animated' conversation not only because it is always lively if it is honest, but also because one uses animation in the broader sense – props, pictures, texts, films and even 'demonstrations.'* (emphasis added)

One more important point is that the teaching of science should be *"mindful of the lively processes of science-making, rather than being an account only of 'finished science' as represented in the textbook, in the handbook and in the standard and often deadly 'demonstration experiment'...."* (Bruner, 1996, p. 127; emphasis added)

IN THE NEMO SCIENCE CENTER

We have often visited the NEMO Science Center, in the Netherlands. Once, when we were watching a group of about 10 year old children in the water worlds exhibition, they tried to make a waterfall and change its direction to reach up to a model house. They were putting sand bags in some different positions and exploring the changes in water flow direction. The kids were so absorbed with their exploration that it was impossible to stop or interrupt them. They may have had instructions from their teachers to do this investigation, but the teachers were not participating. Teachers could not be seen. A bridge between formal and informal learning was evident, and all of the learning process was carried out in a physical and logical and discursive way. Simultaneously, behind the waterfall, there was an exhibition about making bubbles: huge bubbles, small bubbles, ball-shaped and planar bubbles, barrels of bubbles. Some younger children experimented with that issue-based installation. They stepped into the middle of a large pan filled with soapy water, and they were sinking a steel band into the soapy water and trying to pass it around their bodies. In this way they tried to cover their bodies with a large bubble barrel. They were very excited and inspired. They were wondering constantly about the bubbles. They actively explored, experimented, and had many new experiences, which posed new questions; then they started to wonder again and again in an eternal circle. This was learning by doing and not doing to learn. And it was beyond any standardized textbook. They were constructing their own viable and fruitful learning about cohesion forces and potential and kinetic energy of waterfalls.

As Hein showed, beginning in 1997, educational theories can be distinguished by their positions on epistemology and learning theory. Knowledge theories can be represented in a continuum – from Plato's idealism to realism. Many philosophers can be represented on this continuum. Similarly, there are contrasting views on learning theory: theories of passive mind and theories of active mind. The latter extends from Dewey to modern prominent developmental and socioculturalists. Hein shaped this approach as shown in Figure 7.

If we combine passive learning with realism we encounter the usual visits to museums – to admire the treasures and the miracles of the world, accompanied by lectures and texts, which are the prominent means in this approach. If we combine passive learning and idealism, we tend to reproduce behavioral learning based on stimulus-response schema. If we place the priority on active learning and remain anchored to realism, we tend to discover the mysteries of the world as pure scientists do. But if we combine active learning with idealism, we encounter the constructivist

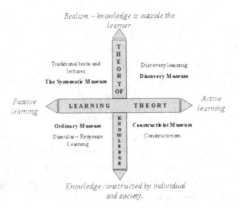

*Figure 7. Modification of the model provided by G. Hein (1997) about different
types of museums related to different educational theories.*

museum in the current sociocultural context. This means that children learn by
doing and achieve a societal kind of knowledge, temporary but viable and flexible,
fruitful, and justified by the current hands-on and minds-on experience.

Vygotsky introduced the concept of internalization. According to this assertion,
development involves a transfer of social interactions to the individual mind.
Vygotsky wrote that "[A]n interpersonal process is transformed into an
intrapersonal one" (1978, p. 57).

Similarly, Lawrence and Valsiner (1993) proposed development as
internalization as in the following assertion. "[W]hat was originally in the
interpersonal (or inter-mental) domain becomes intra-personal (intra-mental) in the
course of development" (p. 151).

Valsiner, following Vygotsky, acts within the constructivist current of thought
and, consequently, rejects conceptualizing internalization as transmission
('exclusive separation'), emphasizing Vygotsky's claim that internalization
involves transformation. All socioculturalists reject theories of development based
on genetically predetermined stages and theories of development that are not
foundationally based on social interaction (Swayer, 2002, p. 294).

Some socioculturalists mention that even Vygotsky emphasized some kind of
dualism, since in his famous work, *Mind in Society*, he considers mind (that is to
say, the individual) and society as components of a whole. Blunden (2010a) traced
the roots of CHAT from Goethe, Hegel, and Marx, and clarifies how Vygotsky
developed his theory of mind, in which the individual and society form a Gestalt.
This study belongs to those which try to overcome a kind of an emphasized dualism
in the work of Vygotsky. Consequently, Blunden, permanently anchored to the
theory, tends to accept Leontiev's approach to Activity Theory, as well as Michael
Cole's cross-cultural research (2005) on the role of context in learning. Blunden
ends with an interdisciplinary concept of activity which overcomes the obstacles of
the dichotomy between the individual and social fields in humanistic thinking.

Swayer (2002, p. 294) focuses on Rogoff's strong advocacy for inseparability and her strong "mutual constitution view:" "The child and the social world are mutually involved to an extent that precludes regarding them as independently definable (p. 28). Rogoff (1997) argues that "the boundary between individual and environment disappears" (p. 267).

However, in empirical practice, Rogoff maintains a three-fold analytic distinction between individual, group, and community, referring to these as "angles," "windows" (1990, p. 26), "lenses," or "planes of analysis" (1997, pp. 267–268).

This may show that a methodological issue has emerged in sociocultural theory. Maybe the positivistic methodology is really so embedded in all research traditions that even socioculturalists familiar with dialectics encounter methodological obstacles, especially what we define as the "unit of analysis in each research." This causes some problems in evaluating the activities and the working-teaching program in the schools. But in informal settings the inseparability works very well and provides fruitful data. Wolf-Michael Roth makes his theoretical assumptions and theses by analyzing concrete episodes from the practical work, which is a pure bottom-up approach that creates activities suitable for children by engaging them in collective curriculum design (Roth & Goulard, 2010).

THE SOCIAL ROLE OF MUSEUMS

The constructivist museum is a reality now, and among other things, it serves social action philosophy. Formal and informal education need to have the potential to empower citizens to make informed decisions in a democratic society (Hein, 2004). Inspired by Hein, if we intend to adopt CHAT as one of the future directions for both science and museum education, we have to consider many domains: exhibition content and purpose, changing museum practices, process ontology, inseparability of individual and society, training of trainers, and social action through process and content.

Museums are strong cultural tools, and their impact on society is gradually advancing. For example, many exhibitions about climate change in natural history museums emphasize on social action. Beyond this focus, museums change their educational practices. They give greater emphasis to community participation and less to marvelous and miraculous exhibitions. And just after the introductory bold experience in the exhibition, they need to design activities–activities that take into consideration the subject, the object, the tools, the rules, the community, and the division of labor. The design requires interdisciplinary working groups with both scientists and practitioners and a new mentality about the societal role of museums and science education. The latter two domains are dialectically complementary, and science museums and science centers are the mediative tools.

The relevant theoretical issues underlying this paper are cultural relativism in relation to human rights, cultural violence, scaffolding in learning science, enculturation in science, and metacognitive procedures in a relativistic approach to values. We argue that the CHAT context is the most appropriate context for science museum education and expands the boundaries of science education in science museums and science centers. Old museums, contemporary museums, environmental centers, local museums, children's museums, interactive parks, zoos, and botanical gardens are some of the settings which provide bold cultural tools for enthusiastic and meaningful learning in the natural sciences. This review paper examines the debate about the way science is re/presented in informal science settings, such as science centers and science museums. In this way the paper aims to bridge the gap between formal and nonformal learning and to develop new science curricula in museum education, thus introducing methodological tools from the field of Cultural Historical Activity Theory (CHAT) linked with current science education literature on the nature of science (NOS) and science, technology, society, and environment (STSE) perspectives.

CHAT AND MUSEUM EDUCATION

CHAT, like all sociocultural theories, accepts and precedes a process ontology. It does not accept separate entities and it does posit the inseparability of the individual and the group (Swayer, 2002). Process ontology and the inseparability of the individual and the group are the first and fundamental bases of sociocultural theory. The latter includes many trends and perspectives that shape a complex web, and we must investigate our position within it. We can put the sociocultural theorist in a web such as that provided by Andy Blunden for the origins of CHAT,[5] or in an interdisciplinary theory of activity (Blunden, 2010a, 2010b). The same way as Blunden argues about the inseparability of the individual and the social domains, we see the inseparability between science and human activity. Blunden, from a critical point of view, poses the substances and premises of a science as follows:

> Thus, the substances, or 'premises' of a science are the conception the researcher has of the ultimate reality underlying the universe of phenomena with which the science is concerned .... naïve realism presumes the existence of matter existing independently of human activity, and obedient to natural laws which are to be investigated. (Blunden, 2010b, p. 4)

The underlying dualism in much sociocultural research can be understood according to the view of scientific subject matter from Kantian skepticism:

> For Kantian skepticism, science deals with a subjective domain of appearances, manifesting things-in-themselves which are beyond perception; so the objects of possible experience are the substance, while 'matter' and 'things-in-themselves' are deemed not to be legitimate objects for science. (Blunden, 2010b, p. 4)

After first getting a physics degree, then a pedagogical degree, and now serving as a science educator, I consider CHAT to be the appropriate physical and logical framework to integrate the natural sciences and humanities. This became very clear as we studied quantum mechanics, relativity, entropy, multiplicity of nature and, recently, ecosystems and sustainability. All those domains emphasize the dialectical complementarity of opposites: for example, linear and cyclic time, Newtonian and relativistic time, before and after in quantum mechanics, and the wave and/or particle nature of light. This leads to humanitarian and or scientific unity into an interdisciplinary concept of activity. Especially, as Blunden wrote:

> Natural scientists can accommodate recourse to the language of activity as a method of description of Nature, while maintaining matter as the substance. But for the human sciences, activity is crucial, for the objects of human life are both constituted and perceived by activity, and this is the key aspect of activity which an interdisciplinary concept of activity must address. (Blunden, 2010b, p. 6)

Prominent science educators see the unity of psychology and sociology in science education and/or propose to reconstruct science education into the intersection of them. Kincheloe and Tobin (2009) write about the positivistic dualism underlying some research:

> A central dimension of our argument is that many of the tenets of positivism are so embedded within Western culture, academia, and the world of education in particular that they are often invisible to researchers and those who consume their research. (p. 513)

The authors continue:

> We have often argued that the epistemology that supports such a dehumanizing and oppressive form of reason is a contemporary form of positivism. (Kincheloe & Tobin, 2009, p. 514)

But while we, as western-educated science educators, were doing science education research, we had a legitimation problem, a methodological dilemma, since we were feeling a lot of bias in our research because we had to explain, to imagine, and to reach conclusions, usually without real empirical data. Tobin expresses this as follows:

> In this positivist context, the only alternatives for educational researchers that exist are either a hard objective stance or a soft subjective position that involves no real empirical data and produces pseudo-knowledges emerging from the imagination of the researcher. (Kincheloe & Tobin, 2009, p. 523)

Returning to science education in science museums and science centers, we find another domain in which CHAT serves as a prominent theory. In recent decades, and since science museums have obtained more and more exhibits embedded in virtual reality, augmented reality, and robots, many researchers have used CHAT to analyze learning and other elements in the evolution of projects. These researchers mainly refer to work about Activity Theory, Human-Computer Interaction, Virtual projects, and building understanding through model building

Using the same research approach, Ogawa et al. (2008) proposed to move forward by advocating for the combination of CHAT with Institutional Theory (IT). Furthermore, they propose to study learning in science museums the same way we study learning in formal organizations. This proposal could reform educational policies, as well as teacher education and in-service training. The

45

authors support the combination of CHAT and IT because they both form a rich theoretical framework that enhances analytic capacity.

Ogawa and fellows also mention that "Both CHAT and IT also promote a view of context that intertwines physical surroundings, organized spaces, daily practices, social relationships, and the meanings that people attribute to the actions they perform" (p. 84). Cole (2005) described context as not just "that which surrounds" some specified unit "in the middle"; rather, he observes, an "act in its context" is "that which weaves together" (pp. 212–215).

Overall, Ogawa and fellows attempting to unite CHAT and IT, rely on two heuristics: Engeström's (1987) model of activity systems and Leavitt's (1965) model of organizations. As a fruitful framework, CHAT provokes the contributions that people make to their learning in combination with all those the sociocultural world offers.

PRACTICAL LEVEL– ANALYSIS OF THE 25 ARTICLES WITH CHAT EXAMPLES: TRAVELING WITH BIRDS[6]

This chapter is based on a meta-analysis of 25 articles for science museum educational programs that were developed under my supervision; most of them have been applied in schools and museums. An annual course on teaching natural sciences in science museums and science centers ended with a conference and a book (Kokkotas & Plakitsi, 2005).

Here we present part of the triangular analysis of an education project entitled "Traveling with Birds: A Holistic Approach through Athens Museums and Open Museum Places."

The educational program focuses on a specific target group: students in Athens schools from 9 to 11 years old. The program was carried out in eight museums. The program's duration was one school year, and it was followed by a summer presentation.

The aims and the objectives were for the students to be able to:
– Know the more important birds of our homeland.
– Become familiar with some of the more important museums and with some open museums.
– Connect education in the natural sciences with arts and culture.
– Become informed about environmental problems, such as the risks to fauna and biodiversity, in combination with the causes behind those threats.
– Realize the societal role of the natural sciences.
– Become proficient in scientific skills and in an evidence-based way of thinking and decision making.
– Practice using their senses during the observation and investigation of the natural world.
– Develop communication skills.
– Develop critical thinking.
– Become familiar with the inseparability of the human and natural worlds.
– Develop commitment to protect the environment.

The holistic approach was developed for a modular educational program having eight modules that correspond to eight museums. School classroom learning communities and museum learning communities can access one or more modules to fit their learning motivations and needs.A group of 16 teachers participated to the museum programs of the eight musem settings choosing a different setting for introductory practice and following their own pathways. All teachers went through the eight different museum settings. The program was taught by me and the local staff of each museum. The museums and the open museum spaces that are proposed by the educational program "Travelling with the Birds" are:

1. *Goulandris Natural History Museum*[7]: Its rich traditional collections of the flora and fauna of Greece make it possible for students to approach birds in their biotopes and habitats, to understand the internal and external interactions of an ecosystem, and to clarify concepts such as biotopes, wetlands, adaptation, food chains, endemic species, migration, predatory birds, etc.

2. *Diomidis Botanical Garden, Athens*[8]: Offers many chances for students to observe important migratory birds in their natural environment.

3. *Aegean Wildlife Hospital – ALKIONI*[9]: Students come in contact with the important work of caring for and protecting birds and their biotopes. They learn how the rehabilitation specialists behave and what they have to do on their own when they find a wounded bird.

4. *Hellenic Wildlife Hospital – EKPAZ*[10]: Students are informed of the dangers that threaten birds and other fauna, recognize the work of volunteers for wounded bird rescue, care, and rehabilitation, and are sensitized to the need to protect ecosystems.

5. *Benaki Museum*[11] *– Archaeological Department*: Students extend their knowledge of birds to archaeological discoveries with representations of birds and come in contact with important samples of ancient Greek art.

6. *Benaki Museum – Department of Popular Art*: Students connect their knowledge of birds with popular culture through various forms of popular art (textile, wood carving, embroidery, etc.).

7. *Byzantine and Christian Museum*[12]: Students are exposed to characteristic samples of Christian art and to the significance of bird symbolism in the Christian tradition.

8. *Numismatic Museum of Athens*[13]: Students connect the images and representations of birds in the coins with the history, art, and culture of countries of origin.

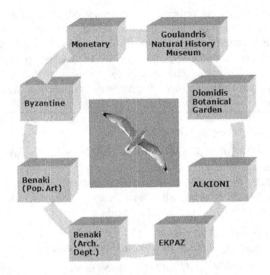

Figure 8. Traveling with birds – Modules.

ACTIVITY ANALYSIS – MODIFICATION OF ENGESTRÖM'S TRIANGULAR ANALYSIS

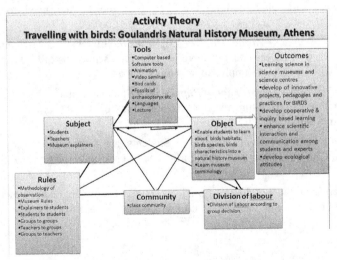

Figure 9. Trianglular analysis of activity in the Goulandris National History Museum, Athens.

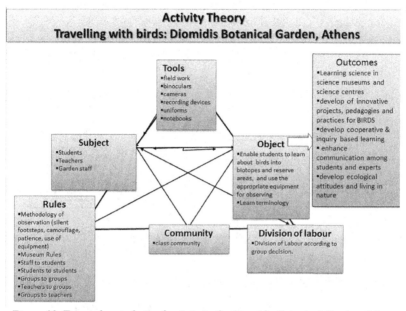

Figure 10. Triangular analysis of activity in the Diomidis Botanical Garden, Athens.

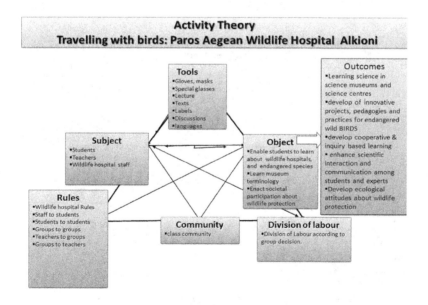

Figure 11. Trianglular analysis of activity in the Paros Aegean Wildlife Hospital, Alkioni.

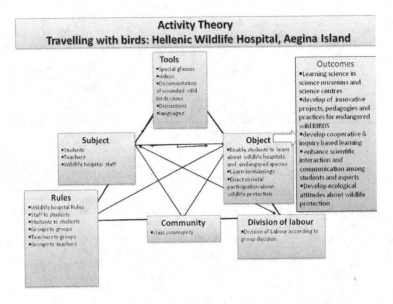

Figure 12. Triangular analysis of activity in the Hellenic Wildlife Hospital, Aegina Island.

Figure 13. Triangular analysis of activity in the Benaki Museum, Greek Ancient Art Department.

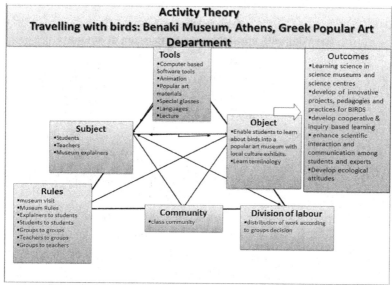

*Figure 14. Triangular analysis of activity in the Benaki Museum,
Greek popular art department.*

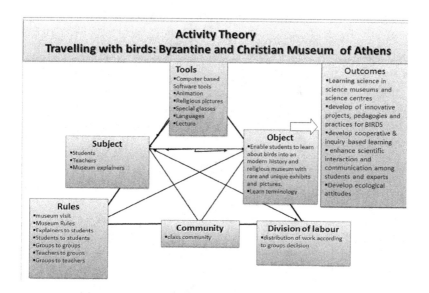

Figure 15. Trianglular analysis of activity in the Byzantine and Christian Museum, Athens.

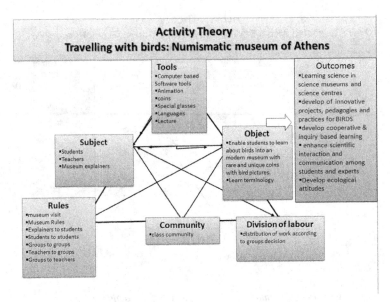

Figure 16. Triangular analysis of activity in the Numismatic Museum, Athens.

ALL SUBTRIANGLES SHARE A COMMON OBJECT

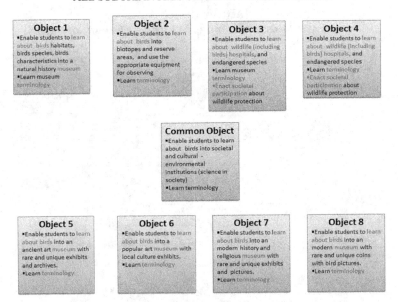

Figure 17. All subactivities share a common object.

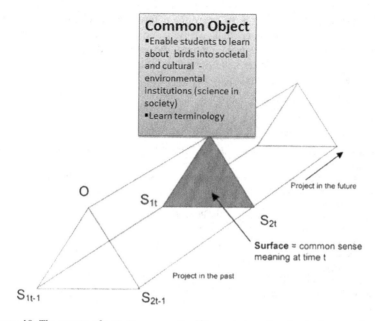

Figure 18. The system of activities moves itself towards transforming its inner actions and operations and represents a new activity in the future.

During the implementation of the educational program, the study of natural sciences is connected with the social and cultural environment in an innovative way. The good organization and preparation of all factors – school teachers, knowledge of object, collaboration between teacher and students in planning, evaluation of field work, an accurate timetable, and achievement of the general and special objectives – resulted in a rich framework that allowed all of us to become "fellow-travelers" with the birds in a journey that was creative and rich in knowledge and experiences.

IN CONCLUSION

Overall, we intend to establish a CHAT perspective in science education in science museums and science centers. CHAT seems to fit into the social role of museums and to overcome the obstacles of positivism in science education and research. At its core, CHAT has the characteristics of multiplicity, dialectics, and unity of differences and in this way enriches the connections between current socioscientific issues represented by issue-based installations in modern science centers.

In this scenario, we can plan each educational program in order to implement sociocultural practices that facilitate the transition from the dichotomies of ordinary scientific thought to a holistic approach to knowing and learning.

If we see an educational program through the prism of CHAT, we can implement changes and ongoing processes, as well as develop the system of activities itself. In other ways, we shall continue to see only a few limited elements of the social activity. We must remember that we will not only design tools but, social participation through social mediation into museums and relevant bold cultural centers of activities. The latter have the potential to transform themselves from tool to object and then to subject and then to community.

NOTES

[1] http://cordis.europa.eu/fp7/sis/
[2] Bourdieu, P., et al. (2000). *Weight of the world: Social suffering in contemporary society*. Stanford University Press.
[3] http://www.aam-us.org/aboutmuseums/abc.cfm#visitors (accessed on 5-8-2010)
[4] http://www.aam-us.org/pubs/mn/MN_MJ06_RelevantMuseum.cfm
[5] http://home.mira.net/~andy/works/origins-chat.htm
[6] Spyratou et al (2005). Supervisor: Katerina Plakitsi
[7] http://www.gnhm.gr/
[8] http://www.diomedes-bg.uoa.gr/start.html
[9] http://www.alkioni.gr/
[10] http://www.ekpazp.gr
[11] http://www.benaki.gr/
[12] http://www.byzantinemuseum.gr/
[13] http://www.nma.gr

REFERENCES

Aikenhead, G. S., & Ogawa, M. (2007). Indigenous knowledge and science revisited. *Cultural Studies of Science Education, 2*, 539–620.

Bakhtin, M. M. (1930/1981). *The dialogic imagination: Four essays*. Ed. Michael Holquist. Trans. Caryl Emerson and Michael Holquist. Austin/London: University of Texas Press.

Barab, S. A., Hay, K. E., & Barnett, M. G. (1999). Virtual solar system project: Building understanding through model building. *Annual Meeting of the American Educational Research Association*. Montreal, Canada: AERA.

Blunden, A. (2010a). *An interdisciplinary theory of activity*. Leiden/Boston: Brill.

Blunden, A. (2010b). *An interdisciplinary concept of activity* (p. 4). Outlines No. 1 2009.

Bruner, J. (1990). *Acts of meaning*. Cambridge, MA: Harvard University Press.

Bruner, J. (1996). *The culture of education*. Cambridge, MA: Harvard University Press.

Cole, M. (2005). Putting culture in the middle. In H. Daniels (Ed.), *An introduction to Vygotsky* (pp. 199–226). New York: Routledge.

Dimopoulos, K., & Koulaidis, V. (2006). School visits to a research center as a form of non-formal science education. *International Journal of Learning, 12*, 65–74.

Engeström, Y. (1987). *Learning by expanding: An activity-theoretical approach to developmental research*. Helsinki: Orienta-Konsultit.

Falk, J. H., & Storksdieck, M. (2005). Using the contextual model of learning to understand visitor learning from a science center exhibition. *Science Education, 89*, 744–778.

Falk, J., & Dierking L. (2000). *Learning from museums. Visitor experiences and the making of meaning*. Walnut Creek, CA: Altamira Press.

Falk, J., & Dierking, L. (1992). *The museum experience*. Washington, DC: Whalesback Books.

Goulart, M.I.M., & Roth, W.-M. (2010). Engaging young children in collective curriculum design. *Cultural Studies of Science Education, 5*, 533–562.

Hein, G. (1997). *The maze and the web: Implications of constructivist theory for visitor studies*. Paper presented at the Visitor Studies Association conference, Alabama.

Hein, G. (1998). *Learning in the museum*. London: Routledge.

Hein, G. (2004). The role of museums in society: Education and social action. Essay published in *Curator: The Museum Journal, 48*(4), 357–363.

Hooper-Greenhile, E. (1992). *Museum and the shaping of knowledge*. London: Routledge.

Karplus, R., Lawson, A.E., Wollman, W., Appel, M., Bernoff, R., Howe, A., Rusch, J.J., & Sullivan, F. (1977). *Science teaching and the development of reasoning – General science*. Berkeley, CA: Lawrence Hall of Science.

Kincheloe, J. L., & Tobin, K. (2009). The much exaggerated death of positivism. *Cultural Studies of Science Education, 4,* 513–528.

Kokkotas, P., & Plakitsi, K. (Eds.). (2005). *Museum education and science education: Theory and praxis*. Athens: Patakis [in Greek].

Koliopoulos, D. (2005). *The didactic approach of a science museum*. Athens: Metexmio [in Greek].

Koster, E. H. (1999). In search of relevance: Science centers as innovators in the evolution of museums. *Daedalus, 28*(3), 277–296.

Lawrence, J.A., & Valsiner, J. (1993). Conceptual roots of internalization: From transmission to transformation. *Human Development 36,* 150–167.

Leavitt, H. J. (1965, March). Applying organizational change in industry: Structural, technological, and humanistic approaches. In G. James (Ed.). *Handbook of organizations*. Chicago: Rand McNally.

Lévy, P. (2001). Collective intelligence. In D. Tend (Ed.), *Reading digital culture* (pp. 253–258). Malden, MA: Blackwell.

Nardi, B. A. (1996). *Context and consciousness: Activity theory and human-computer interaction*. Cambridge, MA: MIT Press.

Ogawa, R. T., Rhiannon C., Loomis, M., & Ball, T. (2008). CHAT-IT: Toward conceptualizing learning in the context of formal organizations. *Educational Researcher, 37*(2), 83–95.

Pedretti, E. (2004). Perspectives on learning through research on critical issues-based science center exhibitions. Published online in Wiley InterScience www.interscience.wiley.com. DOI 10.1002/sce. 20019

Pedretti, E. G., Bencze, L., Hewitt, J., Romkey, L., & Jivraj, A. (2008). Promoting issues-based STSE perspectives in science teacher education: Problems of identity and ideology. *Science & Education, 17,* 941–960.

Rennie, L. J., & Johnston, D. J. (2004). The nature of learning and its implications for research in learning from museums. *Science Education, 88,* S4–S16.

Rogoff, B. (1990). *Apprenticeship in thinking: Cognitive development in social context*. New York: Oxford University Press.

Rogoff, B. (1997). Evaluating development in the process of participation: Theory, methods, and practice building on each other. In E. Amsel & A. Renninger (Eds.), *Change and development* (pp. 265–285). Hillsdale, NJ: Erlbaum.

Roth, W.-M. (Ed.). (2010). *ReUniting sociological and psychological perspectives*. Dordrecht, the Netherlands: Springer.

Roth, W.-M., & McGinn, M. K. (1997). Deinstitutionalizing school science: Implications of a strong view of situated cognition. *Research in Science Education, 27,* 497–513.

Roth, W.-M., & Tobin, K. (2002). *At the elbow of another: Learning to teach by coteaching*. New York: Peter Lang.

Roussou, M. (2004). Learning by doing and learning through play: An exploration of interactivity in virtual environments for children. *ACM computers in entertainment* (Vol. 1, p. 2). New York: ACM Press.

Sawyer, K. R. (2002). Unresolved tensions in sociocultural theory: Analogies with contemporary sociological debates. *Culture & Psychology, 8*(3), 283–305.

Taylor, P. C., & Willison, J. W. (2002, July 11–14). *Complementary epistemologies of science teaching: An integral perspective*. Paper presented at the 33rd annual conference of the Australasian Science Education Research Association, Townsville, Queensland. Available online at: http://pctaylor.smec. curtin.edu.au/publications/asera2002/complementary.html

Vygotsky, L. S. (1978). *Mind and society: The development of higher mental processes*. Cambridge, MA: Harvard University Press.

Katerina Plakitsi
School of Education
University of Ioannina
Greece
kplakits@cc.uoi.gr

KATERINA PLAKITSI

4. RETHINKING THE ROLE OF INFORMATION AND COMMUNICATION TECHNOLOGIES (ICT) IN SCIENCE EDUCATION

The only acceptable point of view appears to be the one that recognizes both sides of reality – the quantitative and the qualitative, the physical and the psychical – as compatible with each other, and can embrace them simultaneously. (Pauli 1995, p. 208)

INTRODUCTION

We agree with the voices that propose a reappraisal of the role of information technology-based research tools in students' development of scientific literacy (van Eijck & Roth, 2007). Van Eijck and Roth argue for "curricula in which the application of IT based research tools nurtures the development of scientific literacy" (2007, p. 235). This argument is along the same lines as many other testimonials to the failure of the promising regeneration of science education using Information and Communication Technologies ICT.

In the same discursive line, "Literature Review in Science Education and the Role of ICT" (Osborne and Hennessy, 2006) describes four arguments for why science education matters: 1. the utilitarian, 2. the economic, 3. the cultural, and 4. the democratic:

The utilitarian: the view that knowledge of science is practical and useful to everyone. However, this view is increasingly questionable in a society where most technologies are no longer accessible by anyone other than an expert.

The economic: the view that we must ensure an adequate supply of scientifically trained individuals to sustain and develop an advanced industrial society.

The cultural argument: the view that science and technology are one of, if not the greatest, achievements of contemporary society, and that a knowledge thereof is an essential prerequisite for the educated individual.

The democratic: the argument that many of the political and moral dilemmas posed by contemporary society are of a scientific nature. Participating in the debate surrounding their resolution requires a knowledge of some aspects of science and technology. Hence, educating the populace in science and technology is an essential requirement to sustain a healthy democratic society. (Osborne & Hennessy, 2006, p. 2)[1]

K. Plakitsi (ed.), Activity Theory in Formal and Informal Science Education, 57–82.

ICTs have a great influence on school teaching and learning, mainly because of the tools they produce. The emphasis on tool-mediated learning may be either more social or more personal, depending on the educational theory that influences a teacher's consciousness.

ICTs dramatically changed the processes of data capture, logging, modeling, and interpreting. Virtual experiments, virtual experiences, and social robots changed the psychological–sociocultural learning environment. Tools achieved such importance that in a way we have often lost the didactical, social, psychological learning orientation and instead focus on the tools themselves.

One crucial point is that ICT tools can be used to gain more time for real discussion, interaction, and interpretation in school science classrooms. Simultaneously, the tools link schools with contemporary science as it is applied in Reggio Emilia ateliers and in various school labs and science centers.

ICTs also play a crucial role during the learning of scientific processes and skills. ICTs support the development of the inquiring mind by facilitating exploration and experimentation and by providing visual feedback in a short time. They also improve self-regulation and cooperative learning among peers by providing motivation and inspiration.

But, as impressive and as important as ICT tools are, there is not a univocal correlation between their use and learning improvement. The effective use of ICTs includes (Osborne & Hennessy, 2006, p. 4)

– ensuring that use is appropriate and 'adds value' to learning activities,
– building on teachers' existing practice and on pupils' prior conceptions,
– structuring activity while offering pupils some responsibility, choice and opportunities for active participation,
– prompting pupils to think about underlying concepts and relationships; creating time for discussion, reasoning, analysis and reflection,
– focusing research tasks and developing skills for finding and critically analyzing information,
– linking ICT use to ongoing teaching and learning activities,
– exploiting the potential of whole class interactive teaching and encouraging pupils to share ideas and findings.

Despite the tremendous impact of ICTs in science learning as well as knowing and learning in general, the current use of ICTs in school science classrooms occurs in a dichotomist way – for example, when the curriculum content imposes the use of ICTs in a linear way. Furthermore, the use of ICTs is separate from the larger pedagogical context and learning activity. Instead of this approach, we see ICTs as inseparable from the instructional objectives in a more free dialectical relationship, with the subject and object of the activity establishing the methodological level of the whole activity. In this way, ICTs are not only the means or methods or lesser materials but the crucial mediative tools between subject and object that form the methodological level of the general entity/process called activity. This is described in Figures 3 and 5 in Chapter 1.

Teachers must be able to mediate the reshaping and transforming of curriculum and pedagogical content knowledge into dynamic outcomes while simultaneously reshaping and transforming goals and pedagogies. The main observable elements are the changes in the activity that create culture and structure–not the use of a dynamic tool, such as ICTs, with limited applications to minor objectives. Fortunately, many theoretical studies, as well as many practitioner studies, have changed teachers' attitudes about and skills related to ICTs.

In England, an independent and non profit organization named future lab (http://www.futurelab.org.uk) reports that the New Opportunities Fund (NOF) scheme for training teachers to use ICT in the classroom appears to have had more success in science than in other subjects (Osborne & Hennessey, 2006, p. 5). Consequently, a dialectical use of ICTs in science education in the CHAT context shifts the evaluation environment towards new dialectical factors in the being and the becoming of the activity system.

Science is one of the major cultural products of western society. Even when it is enriched by ICTs, it fails to engage the interest of students,[2] mainly in Europe and the U.S. Thus, the number of university students in science has gradually decreased in recent years. Below, two tables from the ROSE project (Jenkins & Pell, 2006) show the low impact of science and technology on students.

Some core indicators from Tables 1 and 2 are: Students do not think that science and technology can solve all environmental problems (Table 1), and they do not prefer science to other school subjects (Table 2).

Given this situation, the global community established organizations to carry out large-scale research on science achievement.[3] Two programs, the Trends in International Mathematics and Science Study (TIMSS) and the Programme for International Student Assessment (PISA), have attempted to assess different features of student learning. TIMSS sought to find "what students know," and PISA sought to find out "'what students can do with their knowledge." The data gathered in the TIMSS project related to the intended curriculum (the curriculum specified by the current educational system or other body), the implemented curriculum (the curriculum as taught by teachers, the nature of actual classrooms), and the attained curriculum (what students have learned). The PISA project is concerned with how well 15-year-old students can make use of science knowledge acquired from school and from other sources in situations in everyday life that involve science and technology. Although we have obtained very important data about students' performances in science, both projects followed a context approach and a positivistic methodology. Peter Fensham (2007, p. 168) argues that "both projects achieved very little in providing insights about the factors and conditions that foster better quality science learning (TIMSS) or scientific literacy (PISA). The reports of these two projects give very little sense of what the students are experiencing day by day with their teachers in the science classrooms, and how this can be improved." It may be that the methodology of the two projects, the context approach, fails to gather holistic data using integrated methods of activity system changes observations. Fensham suggests using the culture comparative paradigm, giving an example of Eckstein and Noah's (1991) comparative study of senior secondary examinations as an example of the culture comparative paradigm.[4]

Table 1. Student opinions about the ability of science and technology to solve environmental problems.

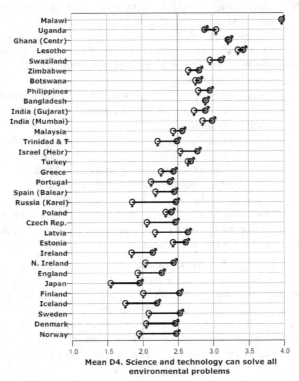

Mean D4. Science and technology can solve all environmental problems

1 = disagree, 4 = agree

So far, all reviews and reports show that most of the curricula are based on the notion that knowledge can be transferred through instruction. This transmission carries predefined and value-free pedagogical content knowledge. This presupposition is totally opposite to real science knowledge acquisition based on cultural-historical activity theory and general sociocultural theory, according to which science knowledge is culturally dependent. However, the demands for changes in new curricula such as 21st Century Science will require teachers to adapt and adopt a different set of pedagogic practices that move students towards scientific literacy (Osborne and Hennessey, 2006). We need teachers to enhance the scientific discourse and methodological skills. We do not need students/citizens who will learn content matters, concepts, or themes, but rather citizens who can negotiate, engage in fruitful discourses on a topic, and who are capable of gathering data and making decisions. Especially, we need new models of student inquiry based on the use of ICTs in a dialectic between the minds-on and hands-on skills of a learning

Table 2. Student interest in school science (Jenkins & Pell, 2006, pp. 68–70).

70

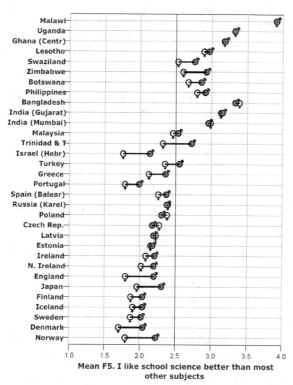

Mean F5. I like school science better than most other subjects

1 = disagree, 4 = agree

subject in correspondence with an inspiring object. In this way, ICTs will be able to transform science education, a result that we have yet failed to achieve.

Thinking of why ICTs have so far failed to transform science education we encounter some sociocultural approaches to ICTs in the context of Human Computer Interaction theory (HCI). As an introduction, we mention Robertson's (2003) assertions concerning the reasons that ICT failed to reform schools. Robertson builds upon Pollitzer (1996) and Carroll (1997) as follows:

HCI researchers, working mainly with the ICT systems and users found in industrial and commercial contexts, have drawn insights from broadly psychological domains ranging from cognitive science in the 1970s and 80s, to a post-Vygotskyian socio-cultural perspective, in the 1990s (Pollitzer, 1996; Carroll, 1997). Areas of particular interest to these researchers have been:
– user-centerd development
– user interfaces
– information and media richness (p. 333)

In recent decades, a major trend for western science educators has been to become familiar with Papert's theories and practices of computing in schools., As a physicist (graduated in 1990s), I have also been trained with programming languages such as Fortran, visual basic, and others. Later, as a teacher, I was trained to teach with LOGO as a means of developing the inquiring mind. Even later, I used all types of ICTs to teach science. Furthermore, as a typical western science educator, I grew up with Piagetian theories, especially Piagetian constructivism versus Papert's constructionism and Papert's discussion with Freire about the future of schools. All that constructivistic background gradually led most science educators in western Europe to other forms of teaching/researching, such as ethnography and other qualitative approaches.

- Robertson concluded by promoting ethnography as an approach that may be: (author's emphasis given)
- rational–based on inadequacy in resourcing, training and/or management
- cultural–based on dissonance between ICT and teachers' professional cultures
- historical–based on long-term views of socio-technological adaption
- gender/cultural–based on female teachers' alienation from ICT's male discourses
- gender/cognitive–based on female (mostly) teachers' learning and interaction style preferences
- cognitive–based on common human preferences for learning and interaction style. (Robertson, 2003, p. 340).

Indeed, many scholars in the last years of the 20th and now in 21st century follow the ethnography paradigm in doing research on HCI. Some prominent HCIs, such as Nardi and Engeström (1999), Nardi (1996), Kaptelinin and Nardi (2006), and Barab et al. (2004), published considerable research that can be integrated into the CHAT context. I think that we have two major metatheories on doing research for educational purposes in the field of HCI. One is to work in the CHAT context in order to see the whole pedagogical context; the other is to apply neuroscience findings in (science) education. Although these two dimensions seem to be different, in fact they have much in common.

ICT theories and applications have been informed by a range of neuroscience data. This combination of ICT and neuroscience influenced a wide range of scientific disciplines. Before continuing, we include in the discussion a short review of neuroscience data that have affected the evolution of science education.

NEUROSCIENCE IN A NUTSHELL[5]

Neuroscience history[6] started from Hippocrates[7] and Plato[8] and continued from the 16th through to 18th centuries, with representatives such as Leonardo da Vinci, Giovanni Battista Odierna, and René Descartes, and then to 1791, when Luigi Galvani published his work on electrical stimulation of frog nerves, and to 1800, when Alessandro Volta invented the wet cell battery. The continuing history of neuroscience continued with some nonscience research, as when in 1836 the novelist, Charles Dickens, described obstructive sleep apnea. In 1838 Robert Remak suggested that the nerve fiber and nerve cell are joined, and then we learned about the "Schwann cell." Then, in 1859, Charles Darwin published *The Origin of Species*. In 1889, William His coined the term *dendrite*, and in 1896 Camillo Golgi discovered the Golgi apparatus. In 1900, Sigmund Freud published *The Interpretation of Dreams*, while Alois Alzheimer (1906) described presenile degeneration. In 1981, Roger Wolcott Sperry, of the California Institute of Technology, was awarded the Nobel Prize for his work on the functions of brain hemispheres. After that and until now, most Nobel Prizes in science have gone to neuroscientists, as, for example, in 1991 when Erwin Neher and Bert Sakmann shared the Nobel Prize for their work on the function of single ion channels, and in Parma, Italy, Giacomo Rizzolatti described (1992) mirror neurons in area F5 of the monkey premotor cortex. Alfred G. Gilman and Martin Rodbell shared the Nobel Prize for their discovery of G-protein-coupled receptors and their role in signal transduction (1994), and Stanley B. Prusiner was awarded the Nobel Prize for the discovery of prions, a new biological principle of infection in 1997. At the beginning of the 21st century, Arvid Carlsson, Paul Greengard, and Eric Kandel shared the Nobel Prize for their discoveries concerning signal transduction in the nervous system, and Linda B. Buck and Richard Axel shared the Nobel Prize for their discoveries about odorant receptors and the organization of the olfactory system.

In spite of impressive scientific discoveries and research, there are still unanswered questions concerning the bridging of neuroscience data with educational practice. Due to the importance of directed attention in cerebral functioning, it is urgent that educational research take into account the emerging data of neuroscience research. The adaptation of neuroscience data provides education an additional scientific confirmation of what is thought to be scientific from an educational point of view. Furthermore, neuroscience can support fields of educational research such as the cybernetics of education, multimedia-based instructional design, applications of simulation in teaching, and the construction of learning environments such as interactive museums. Describing and explaining the change of students' understandings as it occurs in real time is a matter of further research according to Roth (1998).

For neuroscience this means focusing on the pathways of learning and memory processes, especially those concerning the neural parts responsible for the memory. Neurons, by means of electrical signals, carry information through the linking parts of the brain using cell-to-cell junctions called synapses. The ultimate goal of neuroscience is to understand the means by which the neurotransmitters work under certain circumstances. Furthermore, successful teaching is thought to affect brain functions by changing connectivity and potentiation of the synapses (Goswami, 2004).

On the other hand, successful learning is strongly related to successful teaching. In addition, learning cannot simply be conditioned by a teacher's impact. It also depends on the environment that generates the development of the cognitive system in interrelation with the teaching process (Niedderer, 2001). Neuroscience is strongly connected with cultural psychology and then to social interactions, including CHAT. If we adapt and adopt CHAT in learning and teaching science we need to know not only the contribution of pedagogy but also information on how a child's brain functions, how the learning process is taking place in the brain under certain circumstances, and what kind of an educational environment would provide the necessary resources with which the evolution of knowledge and critical thinking could be enhanced. In this way, we propose a combination of CHAT and neuroscience that can provide a coherent explanatory theoretical framework. We shall discuss this issue further later in this chapter.

NEUROSCIENCE DATA INDICATING THE IMPORTANCE OF EXPOSING CHILDREN TO STIMULI

The neocortex is the most recently evolved part of the brain. It consists of about 100 billion brain cells at birth, and its most rapid growth takes place during the first 10 years of life. This is because of the proliferation of dendrites, the branching parts of the brain that receive input through synapses from the linking neurons. This means that due to the lack of environmental input, normal development of the deprived neurons is prevented. In addition, data indicate that passive observation is not enough and that interaction with the environment is clearly necessary. It is at this point that CHAT plays its role. CHAT provides a fruitful context for analyzing all the necessary stimuli inputs and, mainly, the various outputs that usually are totally different from those that teachers had prestructured.

The way that experience strengthens neural connections in the brain was defined by Grossberg (1982; 2005), who has proposed and tested equations describing the basic interaction of the key neural variables involved in learning. One equation of particular importance is his learning equation. It describes changes in the transmitter release rate and identifies factors that modify the synaptic strengths of knobs. Furthermore, as far as learning is concerned, it implies that for information to be stored in long-term memory, two events must occur simultaneously: (1) signals must be received and (2) nodes must receive inputs from other sources in order to cause the nodes' activation.

In science education this means that it is of vital significance for learning instruction to take place along with appropriate visual stimuli such as science models. In particular, these science models should combine further neuroscience data concerning the perception of movement and color in order to become more vivid in children's brains, thus creating permanent ideas.

HOW VISUAL INPUT IS PROCESSED THROUGHOUT DIFFERENT PARTS OF THE BRAIN

As reviewed by Kosslyn and Koenig (1995), there are six brain areas responsible for the ability to recognize an object. In particular, sensory input from the eye's retina passes to the back of the brain and creates electrical activity in the visual buffer located in the occipital lobe. This process develops a spatially organized image in the visual buffer. The attention window (a smaller region in the visual buffer) is also in process. The electrical activity is automatically sent on each side of the brain by means of pathways: two of them run down to the ventral subsystem of the low temporal lobes while two others run down to the dorsal subsystem of the parietal lobes. The dorsal subsystems are responsible for the analysis of factors such as size or location. Electrical activity of the ventral and dorsal subsystems is sent and matched with visual conceptions stored in associative memory (in the hippocampus, the limbic thalamus, and the basal forebrain). Then, depending on whether there is a good match or not, the observed object is categorized in the category to which it belongs (name, sounds it makes, characteristics). In case a good match cannot be achieved, the object remains stored until some further information enables its categorization. When there is an optical information input, individuals create a hypothesis of what it is; when additional input is sought, the working memory (in the prefrontal lobe) starts the procedure of shifting attention in which the eyes focus on a location where the information being sought is likely to be obtained. This procedure is repeated until enough information about the optical stimuli is obtained (Kosslyn & Koenig, 1995).

During learning, the brain prompts us to look around and connect information input with our existing experience, thus generating a combination that will enable our attainment of knowledge.

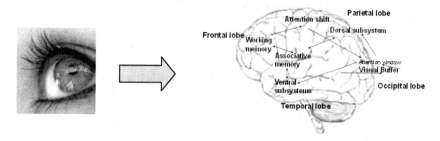

Figure 1. A description of the brain areas that interfere with the visual input.

This means that as far as science education is concerned, when a science model is shown during a science exhibition (either in formal or in informal education) a complex procedure takes place in the child's brain, prompting it to combine existing storage with new stimuli. On a larger scale, this means that with the use of models, the child's brain merges information relative to the subject by either complementing or re-forming pre-existing ideas. It is therefore important that a science model be adequately designed in order to become an appropriate stimulus for activating the brain's function.

EXOGENOUS ATTENTION, COLOR, SPEED PERCEPTION, AND VISUAL MOTION

When designing a science model it would be an error to omit factors strongly related to the brain's learning-memory mechanisms, such as color and speed perception or visual motion.

65

Research results (Fuller & Carrasco, 2006; McKeefry et al., 2007; Pasternak & Greenlee, 2005; Turatto et al., 2006) indicate that attention increases apparent saturation without changing apparent hue, notwithstanding the fact that it improves orientation discrimination for both saturation and hue stimuli. All research data indicate the strong relationship between neuroscience research in the field of visual perception, implying clearly how necessary for science model designers it is that they adopt such information in their efforts to construct an appropriate science model for science teaching, either formal or informal.

CONTRIBUTION OF INFORMATICS IN UNDERSTANDING BRAIN FUNCTIONS

Computational approaches report that internal models are fundamental for under-standing a range of processes such as state estimation, prediction, context estimation, control, and learning (Wolpert & Ghahramani, 2000).

Neuroimaging is based on the assumption that any cognitive task makes particular demands and triggers specific changes in neural activity. Any change of neural activity is strongly related to changes in local blood flow, which enables researchers to measure it either directly (PET) or indirectly (fMRI). Event-Related Potentials (ERPs), on the other hand, are used in order to measure dynamic interactions among mental processes.

Positron emission tomography (PET) is based on the injection of radioactive tracers, and thus is not a suitable procedure for children. The localization of neural functions is feasible due to the accumulation of radiation in brain areas that demonstrate higher levels of blood flow along with higher levels of the tracer.

Functional magnetic resonance imaging (fMRI) allows the localization of brain activity by inserting the participant into a large appropriately designed magnet. The protons of water molecules in neural cells generate the magnetic resonance signal, and fMRI manages to measure it. Depending on blood flow in a particular brain area, the distribution of water in the brain tissue changes as well. The outcome measure in most fMRI studies is the change in BOLD (blood oxygenation level dependent), which enables researchers to measure changes in the oxygenation state of hemoglobin and draw conclusions about neural activity. However, inside the magnet, participants are given headphones in order to shield their ears because it is very noisy. Furthermore, the magnet is claustrophobic, and participants are given a panic button in case they feel uncomfortable. Due to these factors, fMRI doesn't appear suitable for use with children.

However, with the advent of specially adapted coils and less claustrophobic head scanners, fMRI studies are growing in number. Nevertheless, ERP (event-related potential), a widely used technique, can be applied to research involving children. This is because ERPs enable the timing rather than the localization of the focusing neural events. More specifically, sensitive electrodes are placed on the skin of the scalp, and recordings of brain activity are taken. This method of recording the spontaneous and natural rhythms of the brain is called electroencephalography, and ERP refers to systematic deflections in electrical activity that may precede, accompany, or follow experimenter-determined events.

ERP rhythms are thus time-locked to specific events designed to study cognitive function.

According to this method, a head cap (like a swimming cap) with electrodes is placed on the child's scalp while the child watches a video. For visual ERP studies, the video delivers the stimuli; for auditory ERP studies, the linguistic stimuli form a background noise, and the child sits engrossed in a silent cartoon. This way, researchers can record data concerning the latency of the potentials, the amplitude (magnitude) of the various positive and negative changes in neural response, and the distribution of the activity.

The different potentials are called N100, P200, N400, and so on, denoting negative peak at 100 ms, positive peak at 200 ms, and so on. The amplitude and duration of single ERP components such as the P200 increase until age 3 to 4 years (in parallel with synaptic density), and then decrease until puberty. ERP latencies decrease within the first years of life (in parallel with myelinization) and reach adult levels in late childhood. Several studies using ERP techniques have provided strong evidence concerning the cascade of events during neural processing and are sensitive to millisecond differences. The sequence of observed potentials and their amplitude and duration are used to understand the underlying cognitive processes.

Harwitz and Braun's research groups (2004) have undertaken the investigation of network interactions by combining computational neuroscience techniques such as fMRI and PET with functional neuroimaging data. In particular, PET data were used to examine the functional connections between anterior and posterior perisylvian areas during language and language-related production tasks. A large-scale neural modeling approach was illustrated in order to examine visual object processing. The research focused on the centrality of network interactions between brain regions for mediating the performance of sensorimotor tasks. Furthermore, it dealt with the problems raised by using hemodynamic signals as indirect indications of neural activity. The research showed how one could use large-scale neural network modeling to try to relate neural activity to the hemodynamically based data generated by fMRI and PET. A model of object processing for the visual objects was reviewed. This effort concluded that the regions comprising the models corresponded to important nodes along the ventral visual processing pathway from occipital to inferior temporal to prefrontal cortex. What is of great interest is that this simulative model managed to incorporate a specific hypothesis about how visual objects are processed. In addition it enabled researchers to match simulated and experimental data providing support to any hypothesis concerning brain function during visual interpretations. The above model could be a testing tool for evaluating a science model.

Astakhov (2007) reported that based on the Landau principle ($kT\ln(2)\sim1$bit) we cannot have complete knowledge about a complex system such as the human brain. Measurements will require enormous thermodynamic energy to encode everything in bits. Moreover, creating a detailed copy of the complex physical system is impossible due to the fact that measurements will chaotically affect the original and eventually destroy it. However, the author suggests that the next step should be to develop a simulation algorithm for a scale brain network system ($\sim10^{15}$ neurons).

It is therefore vital to recognize the large contribution of informatics in mapping the pathways of learning by understanding the processes of vision, motion, and perception in general. Specifically, informatics takes part not only in the design of learning environments or science models, but also in modeling the human brain network.

Studying complex systems (for example, the knowing and learning brain as science education occurs), we strongly recommend a systemic approach embedded in social and environmental factors. One fruitful systemic approach can be CHAT, especially the interactive systems of activities.

V. S. Ramachandran, in his article "Mirror Neurons and Imitation Learning as the Driving Force behind 'the great leap forward' in Human Evolution," posits that:

> The discovery of mirror neurons in the frontal lobes of monkeys, and their potential relevance to human brain evolution is the single most important 'unreported' (or at least, unpublicized) story of the decade. I predict that mirror neurons will do for psychology what DNA did for biology: they will provide a unifying framework and help explain a host of mental abilities that have hitherto remained mysterious and inaccessible to experiments. (http://www.edge.org/3rd_culture/ramachandran/ramachandran_index.html, p. 1, accessed 4-10-2010)

Ramachandran continues on page 1 with Rizzollati and Arbib's application of their discovery to language evolution. But Ramachandran believes that these neurons will help us understand other important aspects of human evolution as well. Mirror neurons can also enable humans to imitate the movements of others. As Rizzolati has noted, these neurons may also enable us to mime the lip and tongue movements of others which, in turn, could provide the opportunity for language to evolve. Another important piece of the puzzle is Rizzolatti's observation that the ventral premotor area may be a homologue of Broca's area –a brain center associated with the expressive and syntactic aspects of language in humans. These arguments do not in any way negate the idea that there are specialized brain areas for language in humans. We are dealing here with the question of how such areas may have evolved, not whether they exist or not (Ramachandran, p. 1).

The previous discussion adds another point to our argument for the dialectical relationship between CHAT and neuroscience. If we adapt and adopt the arguments about two "big bangs" in human evolution, we can distinguish the first one, which is the invention of highly sophisticated "standardized" multipart tools, tailored clothes, art, music, math, religious beliefs, and perhaps even language, and the second one, which is the emergence of nuclear power, automobiles, air travel, and space travel that occurred after the scientific revolutions by Galileo and Gutenberg. The point is how could two such dramatic changes occur in a shorter period than in the previous periods of human history? One explanation would be the occurrence of genetic changes in the human brain. Another explanation is Ramachandran's:

> … certain critical environmental triggers acted on a brain that had already become big for some other reason and was therefore "pre-adapted" for

68

those cultural innovations that make us uniquely human. (One of the key pre adaptations being mirror neurons.) Inventions like tool use, art, math and even aspects of language may have been invented "accidentally" in one place and then spread very quickly given the human brain's amazing capacity for imitation learning and mind reading using mirror neurons. Perhaps ANY major "innovation" happens because of a fortuitous coincidence of environmental circumstances – usually at a single place and time. But given our species' remarkable propensity for miming, such an invention would tend to spread very quickly through the population – once it emerged. (http://www.edge.org/3rd_culture/ramachandran/ramachandran_p1.html, pp. 4, 5)

From a sociocultural perspective, we note Cole and Hatano's (2007) work, published in the *Handbook of Cultural Psychology*. This work integrates phylogeny, cultural history, and ontogenesis into cultural psychology. Especially, Cole and Hatano say (p. 110) that their intent is to "bring cultural history into the study of human ontogeny without abandoning a commitment to an evolutionary perspective." Cole and Hatano (2007) refer to Vygotsky and Luria's (1930/1993) view of "turning points" in the process of development. A turning point in phylogeny is the use of tools in apes. A turning point in human history is the appearance of labor and symbolic mediation. A turning point in ontogeny is language, which correlates cultural history and phylogeny. In this way the higher psychological functions develop.

This study indicates the need for a Vygotskian perspective on a developmental approach to phylogeny, ontogenesis, and culture–history. Beyond this approach, we adopt Geertz's (1973) famous assertion that there is a co-evolution of phylogeny and culture–history in the development from the early common ancestor of *Homo sapiens* and its nonhuman primate "cousins."

Cole and Hatano take a more critical position on the famous discovery of mirror neurons. They mention the assertion that mirror neurons facilitate, but are not sufficient for, social cognitive development such as imitation.

Changes in morphology and behavior result in changes in the cultural tool kit that result in changes in nutrition, that in turn result in changes in morphology, that in turn result in changes in the cultural tool kit in a never-ending spiral of development. (Cole and Hatano, 2007, p. 131)

The authors also focus on "microgenetic development, that is, cognition and learning at particular phases of ontogenetic development. Finally, the authors argue that into the cultural historical activity theory framework prior knowledge and processing mechanisms are emerging products of the interaction between innate and cultural constraints.

Comparing Cole and Hatano's 2007 article with an older one by the research team at the Laboratory of Comparative Human Cognition (LCHC) on Culture and Cognitive Development, we can find rich arguments about 19th-century anthropology, which was dominated by the following four key assumptions:

- Cognition and culture are aspects of the same phenomenon.
- Cultures are characterized by levels of development.
- Levels of culture (or degrees of civilization) are uniform within societies.
- Change is the result of endogenous mental/social factors. (LCHC, 1983, pp. 296-297, retrieved from http://lchc.ucsd.edu/Pubs/Culture-CognitiveDev.pdf)

In the 20th century the discussion focused on five major issues:

- The adaptive problems of humans because of their common membership in the Homo Sapiens species.
- Many scholars see a single principle of directionality in social history as well as individual biography.
- The unit that serves as the individual is sometimes the individual person and sometimes the individual culture.
- The structure and the content of early experience shape the nature of latter experience. Wordsworth asserted that 'the child is the father to a man,' and he was speaking about anthropology and psychology as well as local folk knowledge.
- At the cultural level, the problem of uniformity is central to discussion of cultural evolution. At the individual level, the problem of uniformity is central to discussion of stages of individual development. The issue of uniformity is central to any theory linking individual behavior to cultural experience, and it is central to all theories of change. (LCHC, 1983, p. 299)

At the intersection between epistemology and psychology, Jean Piaget is the best known representative of cognitive universals. In parallel, according to Piagetian tradition, four factors are considered to contribute to development (LCHC, 1983, p. 300). These are biological factors, equilibration factors, social factors of interpersonal coordination, and factors of educational and cultural transmission (LCHC, 1983, p. 301). This knowledge transfer influenced all levels of academic research and education and was one of the major criteria for successful teaching and learning. Piaget built his epistemology on studies of the mental mechanisms of a child's conceptions of the world, time, object, and so on, and was looking for a general and universal mechanism of reasoning. He defined the well known stages of sensorimotor intelligence, the stage of concrete operations, and the stage of formal operations. Although he did not speak about variability within stages (e.g., that there is an ordering of the acquisition of conservation that begins with the conservation of quantity, then weight, then volume), LCHC (1983, p. 303) supports the view that this is not a prediction derived from Piaget himself but from Piagetians who did not properly integrate the theory. Among others, Piaget greatly influenced science educator researchers such as Rosalind Driver, Andrée Tiberghien, and Jon Ogborn, who are the founding scholars of constructivism in science education. The community of science educators who followed those prominent scholars, as well as Ernst von Glassesferd et al., did basic research for some decades on children's ideas about science. This caused the reformation of modern curricula according to defined levels-standards-benchmarks in order to expand the scope of effective learning and effective schools. In all those approaches culture is considered an independent variable. Comparative studies between urban and rural children and between western and nonwestern societies examined the role of Piagetian factors. Except for biological maturation, all the other factors–the factor of equilibration, of social coordination, and of specific education–remain as plausible sources of differences in cognitive development. Thus, ICTs had a role in supporting teaching and learning in previous forms of knowledge acquisition.

The socialization perspective (LCHC, 1983, p. 310) contains the following propositions: (1) The basic economic activities of a people are constrained by physical ecology; (2) cultures elaborate different kinds of social organization to deal with basic life predicaments; and (3) cultures transmit their acquired wisdom to their children in ways that fit in with a culturally elaborated system of adjustments representing adult patterns of living. This leads to different strategies for survival and problem solving, which accordingly leads to different uses of tools. During the first half of the 20th century Malinowski (1927) undermined the idea that the psychological development is controlled by universal, biologically determined features of the species, thus contradicting Freudian theory. A large number of trends and perspectives in cognitive psychology followed this development, and subsequently, ICTs played a different role in each of them. Psychological differentiation theories, behavioral theories, context-specific approaches, infancy and motor development, perceptual skills, stimulus equivalence and familiarity, classification, memory, communication, the cognitive consequences of literacy, and context interaction are described in the collective work of LCHC (1983, pp. 313–334). All these theories moved from the specific to the general and provided a process model of reasoning, often called a model of information processing. The human mind was considered a computer processor, and the inputs and outputs were studied in more or less open systems.

Later, Vygotsky focused on the relationship of the social and the individual. Vygotsky and his followers worked on the general idea that the social plane is first, and the individual plane follows, a viewpoint that overturned all previous cognitive approaches. In his words the main idea is:

Any function in children's cultural development appears twice, or in two planes. First it appears on the social plane and then on the psychological plane. First it appears between people as an interpsychological category and then within the individual child as an intrapsychological category. This is equally true with regard to the voluntary attention, logical memory, the formation of concepts and the development of the volition. (Vygotsky, 1978, p. 57)

This main idea and later work on the zone of proximal development created a totally new opportunity for the use of ICTs. Then, after the passage of decades, we moved forward to the CHAT context, which is more appropriate for modern and current learning processes in sociocultural theory. In our opinion, CHAT provides a fruitful, open, and systemic context that can be combined with the neuroscience framework to explain learning processes in our civilization. Even cultural-historical activity theory may seem an old theory. In fact, it is considerably broad when compared to other theories, for example, constructivist theory. Being a socio-cultural theory, CHAT interprets a learning situation by paying attention to the broader social system in which the learning is happening and draws interpretations about an individual's thinking and development based on his or her participation in culturally organized activities. An account of learning and development through the lens of constructivist theory, in contrast, is concerned with the individual – and the ways in which sense making happens through the individual's accommodation of experience (Cobb, 1994). Even though Plakitsi's doctoral dissertation is based on the Piagetian approach to the topic of time, it was soon realized that Piaget was influenced by Immanuel Kant and that an individualist approach to development causes many problems when applied in different educational systems. In this context, modern curricula adopted the same predefined standards for all children, and the whole educational system was legitimated to provide success or failure for each child without taking cultural differences, institutional differences, and societal conditions into consideration. Locality was excluded and universality imposed. In contrast, cultural-historical activity theory involves different cultures, different institutional practices, different societal conditions, and keeps the local local, leaving imitation and emulsion and activities to form the different civilizations.

In this context, modern ICTs are the crucial tools that can provide a "turning point" in a child's development if we handle them appropriately. If we only use them to look for children's achievement of predefined standards, the activities lose their meaning, and we wonder why ICTs, which have become the greatest human investment in modern times, failed to reform schools. Fortunately, scholars, working from a sociocultural perspective, study expanded concepts, such as distributed cognition, that include not only people and artifacts, but also ICTs. For example, Shaffer and Clinton (2006) introduce a new category of tool, which they call *toolforthoughts*. They argue that media, such as video games, word processors, and analytical tools, create new skills and habits of mind, and also shift the focus from reading and writing the printed word to multimodal literacy.

Recently, sociocultural theory has been considered in the design of online, e-learning, and virtual learning technologies; simultaneously, science museums and science centers have increased their use of virtual reality, augmented reality, and robotic learning technologies.

One application of those technologies is the Polymechanon Park (www.polymechanon.gr) in Greece. This interactive science park includes a Wobble board and full-body games (Figures 2a, 2b, 2c). The Wobble board game involves the interaction between a composite weight on a 5 x 5-meter floor and a virtual board balanced on its center (Figure 2a). Changes in the measure and location of the composite weight result in displacements of the virtual Wobble board. Up to 12 players need to collaboratively move on the floor board in order to move the virtual board; the goal is to displace a number of balls so that they drop into a number of fixed holes on the virtual floor. The players need to gain control of how the virtual board moves so that they can try and get the balls rolling into the holes. To do that they need to negotiate how they will move themselves on the board. The ideas of forces, balance, weight, location, and direction are embedded in the game, which has various parameters such as simulated friction of the balls on the virtual floor, depth of holes, and collision rules (http://www.makebelieve.gr/mb/www/en/port-folio/museums-culture/ 62-polymechanon.html).

The *Virtual Playground* is an engaging simulation environment for children between 8 and 12 years old and is used for research in evaluating *interactivity* in *virtual environments* for *learning*. Children wear 3D glasses and use a 3D computer mouse (called a "wand") to design a virtual playground. In this playground, the child user assumes the role of a designer who must carry out tasks such as planning the layout of the playground by modifying, resizing, and placing its various elements. These tasks require solving mathematical fraction problems.

Figure 2. (a) The Wobble Board. (b) The meteorite full-body game.
(c) One of the robots guided by children.

Figure 3. The virtual playground.

Another important trend introduced by the European Committee is the networks of excellence, which have barely been influenced by sociocultural theory. Some of them are very important for linking science – ICTs and society, for example, EPOCH European Network of Excellence in Cultural Heritage. This is the European Research Network on Excellence in Processing Open Cultural Heritage. EPOCH was an EU-funded network of about 100 European cultural institutions that joined their efforts to improve the quality and effectiveness of the use of Information and Communication Technology for Cultural Heritage.

This new research area has been less influenced by sociocultural theory so far. According to our approach, many new research approaches will be needed in this specific domain in order to integrate ICT tools, which are bold cultural tools, with the ongoing, endless, developing system of activities. In other instances, the United States and Europe will spend a great deal of money for poor educational effectiveness.

Being optimists, we hope that the advent of this educational technology and its widespread use in schools will provide the dynamism needed to reshape the curriculum of science, especially the pedagogy of science. Easy access to a large body of current and old data gives children and adults the opportunity to learn about science in its development. This is totally different from the previous approach. Is seems as if the global community is one whole body, united in spite of its race, gender, religious, and political differences. This is an emerging way of introducing universalism into the dialectic. All the different systems of activities come together to make one system that interacts with all its components.

So far, in schools at the microlevel, in all national curricula, and in UNESCO priorities, there are great expectations about the potential of ICTs to transform teaching and learning. ICTs can facilitate learning about scientific theories but, even more, learning about scientific processes through interactive investigations using a collective or cooperative approach. The range of ICT applications in science education in schools extends from pure science discoveries (demonstrations using lasers, 3D images, etc.) to virtual playgrounds and virtual experiments, to simple spread sheets, graphing tools, etc. Important teaching and learning methods use new information systems such as Encarta, Wiki, Internet, Intranet, interactive whiteboards, and sensors. Sensors can be used for collecting data, for example, for monitoring weather. A data logging system can support many curriculum areas, such as life and physical processes, materials, and substantial scientific inquiry itself. Some educational CD-roms offer a virtual microscope to introduce children to the use and images of real microscopes. On a larger scale, other educational CD-roms offer a virtual planetarium to introduce children to the use of a real planetarium. In this way scientific knowledge can become public, and early learners can enjoy current scientific discoveries as a means of achieving scientific literacy. Rotated labs can serve this approach as well. In these labs, children change roles and activities and demonstrate and discuss their own knowledge acquisition process. Finally, science is in the same teaching and learning system as language, and this learning acquisition has accomplished a major transformation. Children and adults read, write, and generally communicate using iPhones, iPads, eBooks, YouTube, Facebook, Twitter, Wikipedia, and E Ink. In a volume entitled *Science, Learning, Identity,* Roth and Tobin (2007) explore the belief that beyond a human being's core identity, built on autobiographical narratives of self, there is an ongoing process of creating new identities in relationships with the others. Different and multiple communities create and re-create different identities. Using these modern tools facilitates this process.

But when we use bold ICTs we can lose sight of the specific goals of teaching and learning. Any learner must at least be aware of the objectives of the activity. The use of prominent ICT tools without a dialectical relationship between the learning subject and the object is only a show and does not lead to meaningful learning. Only when tools mediate subject-object authentic interaction into a meaningful activity system can meaningful learning occur. In this scenario, any sociocultural integration of ICTs into the whole activity system will face many obstacles as some predefined concepts, formulas, and other typical forms of Western science. In contrast, scientifically and technologically literate citizens of the 21st century must handle and manipulate different cultures in a dialectical unity. Especially, curriculum agendas are shifting towards multiplicity, dialectics, and interaction with open outcomes. Thus, we need more context research studies in order to be able to develop the societal mediative role of ICTs.

The failure of ICTs to reform schools, in spite of the considerable investment in them, is related to the unsolved problem of the "learning paradox." In 1985, Carl Bereiter published a paper entitled "Toward a Solution of the Learning Paradox." Bereiter adopted the problem as it had been formulated by Fodor 10 years earlier during a meeting between Piaget and Chomsky at Royaumont in 1975. The learning paradox was presented by Fodor as follows:

> ... it is never possible to learn a richer logic on the basis of a weaker logic, if what you mean by learning is hypothesis formation and confirmation .... There literally isn't such a thing as the notion of learning a conceptual system richer than the one that one already has; we simply have no idea of what it would be like to get from a conceptually impoverished to a conceptually richer system by anything like a process of learning. (Fodor, 1980, pp. 148–149)

Fodor and Bereiter accepted that hypothesis formation is an inductive process. Many constructivists such as Ernst von Glasersfeld[9] proposed a solution to the problem of the learning paradox by adopting the Piagetian scheme theory; however, the problem still exists. Recently, van Eijck and Roth (2007), in their paper "Rethinking the Role of Information Technology-Based Research Tools in Students' Development of Scientific Literacy," referred to the problem of the learning paradox and proposed that CHAT could overcome this problem. They argue that:

> expecting students to already know the answer to the problem or question before they conduct research is, indeed, an impossible task (or, as a variant, doing research when students actually already know the answer is a meaning-less task). (Eijck and Roth, 2007, p. 235)

CHAT can overcome all those constraints and provide a rich, open, fruitful context where students can solve the authentic problems that they are involved in. The ICTs are important cultural mediative tools in CHAT. They are not objectives but tools, and the objectives are not strict; they lead to various and maybe unexpected outputs, often more important than the predefined teaching goals. Maybe we can consider that the scheme theory was a forerunner of CHAT systemic analysis.

As Kaptelinin and Nardi state the issue in their book *Acting with Technology: Activity Theory and Interaction Design*, "Activity theory fits the general trend in interaction design toward moving out from the computer as the focus of interest to understanding technology as part of the larger scope of human activities" (p. 5). Since ICTs are embedded in science school practices, we have to adopt the unique potential of CHAT to relate consciousness to activity. In CHAT context consciousness is realized by what we do in everyday practical activity. And this everyday practical school activity includes ICTs in educational systems all over the world. These practices are informed by human knowledge, skills, and values, so they are context oriented, societal, and culturally enriched.

Focusing on the users, who usually are children, we must consider the work of Vygotsky on art and life (1925/1971).[10] He proposes that some people believe that art is the supreme human activity, while others consider it nothing but leisure and fun, and that it is dangerous to give children very attractive graphics and games without a coherent pedagogical context. And in relation to consciousness, Vygotsky states that we have to use ICTs in order to mediate between lower and upper psychological functions. So, ICT is transformed from a material or hands-on tool to a bold mind tool for thinking and learning. We must help children overcome the tendency to focus only on graphical impressions – the tool – a tendency which causes them to lose the significance of the whole activity. We must abandon the typical western technocentered perspective and move towards an understanding of technology as part of human societal activity and interaction with the environment. WebQuest is another example of children handling the control of using ICTs. They solve real problems, gather materials, and create their own websites. Another important role of ICTs is to mediate the global goal of "education for all."

An example from the next chapter concerns the development of an ICT environmental education package entitled: "Environment–Forest protection." The general scope is to develop positive attitudes concerning forest protection among students, even from the early grades. Moreover, very young children (5–9 years old) have some limited opportunities to carry out field work and to experience interactions in an ecosystem. ICT applications can mediate children's learning and acting. For this project, we developed software that modeled a large number of interactions in and out of a national forest in Greece (Map 2).

Figure 4. A map with the national forests in Greece.

Using the map in Figure 4 and images in Figures 5, 6, along with a student's workbook, a companion, and many links to materials from different NGOs, students go through a large number of modular, exploratory, discursive, and cooperative operations and actions and activities. The whole project is represented in Figure 7.

Figure 5. Representation of some forest animals before and after a fire that is common in Greek national forests, especially during the summer.

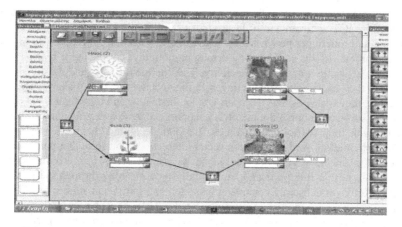

Figure 6. Representation of the energy flow with the software Models Creator.

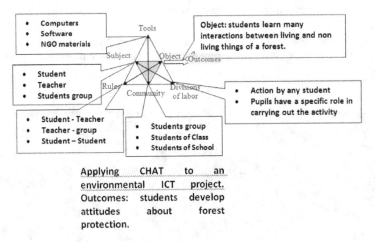

Applying CHAT to an environmental ICT project. Outcomes: students develop attitudes about forest protection.

Figures 7. Triangural analysis of the project "Environment-Forest protection."

CODA

There are three basic reasons for rethinking the role of ICTs in science education: epistemological reasons, learning reasons, and social reasons. ICTs have changed the way scientists act and do their research. Learning enterprises have been totally changed by ICT applications. ICTs are part of children's everyday life. Instead of defining and redefining new roles for teachers and students, we propose to use the CHAT context. In this context the learning community, consisting of teachers and students, acts as the subject of the activity. They define the rules and the division of labor and they choose the tools; acting with this technology, they make reality, they make life, they make society. Teachers are not the unique authority figures but one of the many subjects authorized to act. ICTs support cooperative learning and collective activity with new scaffolding methods. Beyond designing the tools, we design social participation. And through the social interactions that form that participation we develop the new citizen of the world who must act in it. Activism plays a central role in this enterprise and has a totally new meaning through the prism of CHAT. Activism is usually considered as a context, but recent studies put it in an analytic category for theorizing learning (Roth, 2010). Thus the discussion of ICT applications in science/environmental education is enriched by the emergence of CHAT.

NOTES

[1] www.futurelab.org.uk/research/lit_reviews.htm
[2] www.oecd.org/.../0,3746,en_2649_34319_41689640_1_1_1_1,00.html
[3] http://www.pisa.oecd.org, http://timss.bc.edu
[4] On p. 168, Fensham writes that Eckstein and Noah used national documents to see how each country framed its discourse about these examinations, visited each country, and worked with local scholars

and the range of participants in the examining procedures. On site, they could observe and conduct openly structured interviews to elucidate what they saw and heard. In these ways, they gathered a web of data in each country that led to their strikingly vivid set of national narratives, which describe the examinations and their systemic settings through two students from very different family backgrounds who undertake these examinations. For more details, see Eckstein and Noah (1991). *Capable* to gather data and make decisions based on them.

[5]   This text is a rewriting of the previous study of ours: Plakitsi, K., & Dosi, V. (2008). The importance of neuroscience research in science education. *Proceedings IMSCI'08, The 2nd International Multi-Conference on Society, Cybernetics and Informatics*, Orlando, Florida, USA. Volume IV, Post-Conference Issue, Edited by: F. Welsh, F. Malpica, A. Termante, J.V. Carrasquero, A. Oropeza. Awarded as the best paper of the session.

[6]   For a brief list of neuroscience history: http://faculty.washington.edu/chudler/hist.html

[7]   460–379 BC.

[8]   387 BC.

[9]   See invited paper "Scheme theory as a key to the learning paradox," by Ernst von Glasersfeld, presented at the 15th Advanced Course, Archives Jean Piaget, Geneva, September 20–24, 1998.

[10] http://www.marxists.org/archive/vygotsky/works/1925/index.htm

REFERENCES

Astakhov, V. (2007). Continuum of consciousness: Mind uploading and resurrection of human consciousness. Is there a place for physics, neuroscience and computers? *Toward a Science of Consciousness*, 2008 April 8–12, 2008.

Barab, S. A., Thomas, M. K., Dodge, T., Squire, K., & Newell, M. (2004). Critical design ethnography: Designing for change. *Anthropology and Education Quarterly*, 35(2), 254–268.

Bereiter, C. (1985). Towards a solution of the learning paradox. *Review of Educational Research, 55*, 201–226.

Bisley, J. W. & Pasternak, T. (2000). The multiple roles of visual cortical areas Mt/MST in remembering the direction of visual motion. *Cerebral Cortex, 10*, 1053–1065.

Carroll, J.M. (1997). Human-computer interaction: Psychology as a science of design. *International Journal of Human-Computer Studies 46*, 501–522.

Cobb. P. (1994). Where is the mind? A coordination of sociocultural and cognitive constructivist perspectives. In C. T. Fosnot (Ed.), *Constructivism: Theory, perspectives, and practice* (pp. 34–52). New York: Teachers College Press.

Cole, M., & Hatano, G. (2007). Cultural-historical activity theory: Integrating phylogeny, cultural history, and ontogenesis in cultural psychology. In S. Kitayama & D. Cohen (Eds.), *Handbook of cultural psychology*, pp. 109–135.

Eckstein, M. A., & Noah, H. J. (1991). *Secondary school examinations: International perspectives on policies and practice*. New Haven and London: Yale University Press.

Fensham, P. J. (2007). Context or culture: Can Timss and Pisa teach us about what determines educational achievement in science? *Internationalisation and Globalisation in Mathematics and Science Education*, Section 2, 151–172.

Fodor, J. (1980). *Language and learning* (M. Piatelli-Palmerini, Ed., pp. 143–149). Cambridge, MA: Harvard University Press.

Fuller, St., & Carrasco, M. (2006). Exogenous attention and color perception: Performance and appearance of saturation and hue. *Vision Research, 46*, 4032–4047.

Geertz, C. (1973). *The interpretation of cultures: Selected essays* (p. 476). New York: Basic.

Goswami, U. (2004). Neuroscience and education. *British Journal of Educational Psychology, 74*, 1–14.

Grossberg, S. (1982). *Studies of mind and brain*. Dordrecht, the Netherlands: D.Reidel.

Grossberg, S. (2005). Linking attention to learning, expectation, competition, and conciousness. In L. Itti, G. Rees, & J. Tsolsos (Eds.), *Neurobiology of attention*. San Diego, CA: Elsevier.

Harwitz, B., & Braun A. R. (2004). Brain network interactions in auditory, visual and linguistic processing. *Brain and Language, 89*, 377–384.

Jenkins, E. W., & Pell, R. G. (2006). *The relevance of Science Education Project (ROSE) in England: A summary of findings.* Leeds: Centre for Science and Mathematics Education, University of Leeds.

Kaptelinin, V., & Nardi, B. (2006). *Acting with technology: Activity theory and interaction design.* Cambridge: MIT Press.

Kosslyn, S. M., & Koenig, O. (1995). *Wet mind: The new cognitive neuroscience.* New York: The Free Press.

Malinowski, B. (1927). *The father in primitive psychology.* New York: Norton.

McKeefry, D. J., et al. (2007). Speed selectivity in visual short term memory for motion. *Vision Research, 47*, 2418–2425.

Nardi, B., & Engeström, Y. (1999). A web on the wind: The structure of invisible work. A special issue of *The Journal of Computer-supported Cooperative Work, 8*(1–8). Introduction to the issue.

Nardi, B. (Ed.). (1996). *Context and consciousness: Activity theory and human-computer interaction.* Cambridge: MIT Press.

Niedderer, H. (2001). Physics learning as cognitive development. In R. H. Evans, A. M. Andersen, & H. Sørensen (Eds.), *Bridging research methodology and research aims* (pp. 397–414). Student and Faculty Contributions from the 5th ESERA Summerschool in Gilleleje, Denmark. The Danish University of Education.

Osborne, J., & Hennesy, S. (2006). *Literature review in science education and the role of ICT: Promise, problems and future directions.* Report 6: Futurelab Series. Retrieved from www.futurelab.org.uk/research/lit_reviews.htm

Papert, S. (1980). *Mindstorms: Children, computers and powerful ideas.* Harvester Press.

Pasternak, T., & Greenlee, M. (2005). Working memory in primate sensory systems. *Nature Reviews Neuroscience, 6*, 97–107.

Pauli, W. (1995). *The influence of archetypal ideas on the scientific theories of Kepler. The interpretation of nature and the psyche* (p. 208). London: Routledge and Kegan Paul.

Pollitzer, E. (1996). The evolving partnership between cognitive science and HCI. *International Journal of Human-Computer Studies, 44*, 731–741.

Ramachandran, V. S. Mirror neurons and imitation as the driving force behind "the great leap forward" in human evolution. In *EDGE: The third culture.* Retrieved from http://www.edge.org/3rd_culture/ramachandran/ramachandran_p1.html.

Robertson, J. W. (2003). Stepping out of the box: Rethinking the failure of ICT to transform schools. *Journal of Educational Change, 4*, 323–344.

Roth, W.-M (1998). Learning process studies: Examples from physics. *International Journal of Science Education, 20*(9), 1019–1024.

Roth, W. M. (2010). Activism: A category for theorizing learning. *Canadian Journal of Science, Mathematics and Technology Education, 10*(3), 278–291.

Shaffer, D. W., & Clinton, K. A. (2006). Toolforthoughts: Reexamining thinking in the digital age. *Mind, Culture, and Activity, 13*(4), 283–300.

Turatto, M., et al. (2007). Attention makes moving objects be perceived to move faster. *Vision Research, 47*, 166–178.

Van Eijck, M., & Roth, W. M. (2007, June). Rethinking the role of information technology-based research tools in students' development of scientific literacy. *Journal of Science Education and Technology, 16*(3), 2007.

Van Eijck, M., & Roth, W. M. (2007). Rethinking the role of information technology-based research tools in students' development of scientific literacy. *Journal of Science Education and Technology, 16*(3), 225–238.

Vygotsky, L. S. (1978). *Mind in society: The development of higher psychological processes* (M. Cole, V. John-Steiner, S. Scribner, & E. Souberman, Eds.). Cambridge, MA: Harvard University Press.

Vygotsky, L. S. (1999). *The collected works of L. S. Vygotsky, Vol. 6: Scientific legacy* (M. J. Hall, Trans, R. W. Reiber, Ed.) New York: Plenum Press.

Vygotsky, L., & Luria, A. (1993). *Studies on the history of behavior. Ape, primitive, and child.* Hillsdale, NJ: Erlbaum. (Original work published 1930.)

Wolpert, D. M., & Ghahramani, Z. (2000). Computational principles of movement neuroscience. *Nature Neuroscience, 3*(Suppl.), 1212–1217.

Katerina Plakitsi
School of Education
University of Ioannina
Greece
kplakits@cc.uoi.gr

KATERINA PLAKITSI, ELENI KOLOKOURI,
EFTYCHIA NANNI, EFTHYMIS STAMOULIS
AND XARIKLEIA THEODORAKI

5. CHAT IN DEVELOPING NEW ENVIRONMENTAL SCIENCE CURRICULA, SCHOOL TEXTBOOKS, AND WEB-BASED APPLICATIONS FOR THE FIRST GRADES

INTRODUCTION

This paper describes some issues in the design, development, and production of a number of teaching and learning materials and their implications for environmental science curricula, school science textbooks, and web-based/ICT applications.

Our research team is experienced (Plakitsi, 2008, 2010) in writing school science textbooks and in constructing web-based applications for the first grades from an inquiring and a sociocognitive perspective. Even though severely constrained by the Greek Ministry of Education[1] and bounded by a highly structured curriculum, we attempted to develop materials that had open space for an agency/structure teaching schema (Goulart & Roth, 2010). Furthermore, we tried to use many cultural tools in order to engage young students in a collective curriculum design. This approach aims to influence formal and informal learning, indoor and outdoor activities, and to act beyond textbook or software limitations. After developing the textbooks and the software, we moved a major step forward, by doing a meta-analysis of the materials using CHAT. In any project included in the school environmental science textbooks or web-based applications, every system of activity has been linked with one or two related ones. Multiple links can lead to another system of activity, and so on. The interactive systems of activities can share the same object and lead to outcomes that are part of scientific literacy and citizenship. In our opinion, in the early grades the previous scope can be achieved through collective curriculum design processes. Fundamentally, in each activity system, students practice basic scientific skills, for example, observation, classification, measurement, and communication. Some scientific attitudes and values underlie the rules of the learning community (class, group, individuals), and also in the division of labor. The children's previous experiential knowledge becomes the base for the teaching/learning process. Outcomes can be the gradual construction of some basic concepts, so the school transformation of the content knowledge emerges more from inquiry-based learning as well as collateral learning (agency/structure schema) and less as a result from a target-obstacle procedure (constructivism).

K. Plakitsi (ed.), Activity Theory in Formal and Informal Science Education, 83–110.
© 2013 Sense Publishers. All rights reserved.

Level 1: NOS Studies

As a physicist and pedagogist and having served as a science educator at the university, I created a research team. We started the pedagogical/didactic transformation of the scientific content knowledge by adapting some modern aspects of Nature of Science (NOS). In parallel, we taught science in science museums and science centers (see Chapter 3) and also used ICT as a bold mediative tool in activity systems (see Chapter 4). Then we expanded our science educational research to sociocultural approaches such as CHAT. These efforts are represented by three widely known communities:[2] the IHPST, the EARLI (Piliouras, Plakitsi, & Kokkotas, 2007), the ISCAR community (Plakitsi & Kokkotas, 2005), as well as the IOSTE [and NAESA] (Plakitsi & Kokkotas, 2006).

Our first study on NOS (Plakitsi & Kokkotas, 2004) was a review of the modern aspects of the nature of science, which was applied in the context of a Comenius 2.1. Project entitled The MAProject. The basic aim of this program is to encourage teachers to reconceptualize their views about the nature of science, as well as the nature of learning and the nature of teaching. It aims to shift teachers' epistemological views of the nature of science from traditional to contemporary perspectives (Bartholomew, Osborne, & Ratcliffe, 2004; McComas, Clough, & Almazroa, 1998). The basic dimensions are listed in Table 1.

Initially, we reviewed teachers' views of the nature of science (Plakitsi & Kokkotas, 2004, 2006). We found that teachers believe the common position that science, an expression of human creativity and civilization, cannot be established as a meaningful activity unless it transcends the field of quantitative calculation. They also avoid any interpretive-philosophical mediation about the nature of science.

Then our general aim was to challenge in-service primary teachers to "see" science, education, and knowledge in their cultural context. In order to achieve our targets we worked on three levels.

- At the first level, we tried to enhance teachers' nature of science webs, adding more social and cultural aspects.
- At the second level we tried to modify teachers' opinions about the nature of learning and the nature of teaching from a viewpoint of teachers as dispensers of the one right knowledge ("the truth") to facilitators of students' own learning and to a viewpoint of learning that includes the development of reasoning and collaborative and discursive skills.
- At the third level, armed with some cases from the history of science, e.g., the case of falling bodies from the Aristotelian or the Galilean point of view, we tried to help teachers change their views about the nature of scientific knowledge and also about the different contexts and procedures from which that knowledge is produced by scientists.

Table 1. Shifting teachers' views from traditional to contemporary perspectives on the nature of science, nature of learning, and nature of teaching.[3]

Teachers' Knowledge and Understanding of the Nature of Science		
One-dimensional–empiricist view of NOS	⟶	Pluralistic view of NOS
Teachers' views of their own role		
Dispenser of knowledge	⟶	Facilitator of learning
Teachers' use of discourse		
Closed and authoritative	⟶	Open and dialogic
Teachers' views of learning goals		
Limited to knowledge gains	⟶	Include the development of reasoning, collaborative, and discursive skills
Teachers' views about the nature of classroom activities		
Student activities are contrived and inauthentic	⟶	Activities are authentic and owned by students

The research focused on teachers' science education. The sample consisted of 100 teachers who teach science in primary schools and also early childhood education. All the teachers answered a questionnaire, which was a translated part of Aikenhead's 1989, instrument[4] (VOSTS). Furthermore, 50 of the teachers, randomly selected, were interviewed, so we could better study their NOS webs. Teachers took a semester course about the nature of science. During the course – and in order to challenge teachers to understand that science evolution is not a linear process where new knowledge is added bit by bit to the old knowledge – teachers studied and discussed scientific revolutions in the history of science. Furthermore, in order to change teachers' beliefs that there is only one right version of scientific knowledge, we used the following procedure: teachers were divided into two groups; the first group had to study and explain phenomena concerning, for example, falling bodies from the Aristotelian point of view; the second group had to study and explain phenomena concerning falling bodies from the Galilean point of view. Finally, the two groups participated in a public debate where they had to defend their arguments (Kokkotas & Vlachos, 1997; Kuhn, 1993; Plakitsi & Kokkotas, 2005). In order to challenge teachers to expand their nature of science web to include more cultural aspects, they were obliged to study resources and then argue about how scientists from different cultures and/or at different times offer different explanations and conclusions when observing the same phenomena. Twenty of them were obliged to teach the topic of falling bodies to 12-year-old pupils using multiple explanations of the same phenomenon. In this way they enhanced argumentation in their classrooms. Furthermore, they had to teach falling bodies in alternative learning environments such as science museums. They also made mobile materials to teach the topic of falling bodies, such as simple pendulums and incline planes. For the evaluation we used nature of science webs, which were gradually being modified after each stage of the program. We also used queries, but we focused on interviewing. Data collection was done by a questionnaire, some interviews, many audiotapes, and student drawings and videotapes.

RESULTS

Part I. Teachers Views of Nature of Science Extracted from the Interviews (Plakitsi & Kokkotas, 2004)

According to Machamer's (1998) analysis and after studying teachers' nature of science webs (Figures 1–3), we could see a gradual shift from one-dimensional to more pluralistic views. In Figure 1, teachers described science more quantitatively and focused on the mathematical tools that science uses and also scientific measurements. They supported the view that science is creative, but they related this creativity more to the processes and skills of the scientific method, and less to any clever prediction or image.

As we can see in Figure 2, when applying our research program, teachers expanded their nature of science web to include more cultural aspects. They referred to common values during scientific research and to the naturalistic nature of science, while they argued about multiple methods in making science instead of their previous references to only one right and valued scientific method.

Finally, at the end of the research program (Figure 3) teachers in the sample expanded their nature of science web to society and culture much more than in the previous stages. They argued that science includes criticism, that it is a pure social activity, and that it requires communication and publicity. Concerning understanding and comprehension of science, they mainly believe that understanding results from scientific curiosity, which emerges from scientific creativity (see McComas et al., 1998). Most of the teachers said that science gives answers to all that happens around us. They believe that experimentation is crucial for scientific evolution. They also believe that the schema of hypothesis-observation-experiment–conclusion

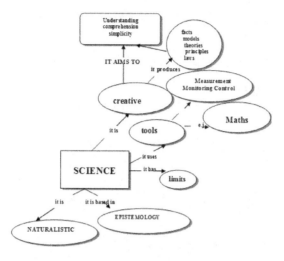

Figure 1. Teachers NOS webs elaborated in one representative example. (Collapsed categories of answers – from those with the highest frequency – are presented in a total NOS web.)

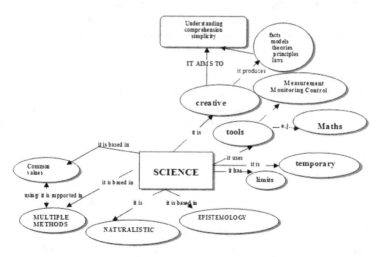

Figure 2. Teachers' NOS webs elaborated in one representative example. (Collapsed categories of answers-from those with the highest frequency-are presented in a total NOS web.)

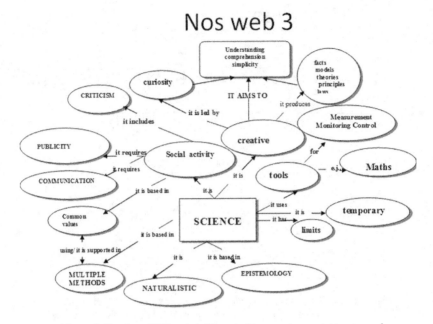

Figure 3. Teachers NOS webs elaborated in one representative example.

comprises the core of scientific method. Furthermore, they believe that data control, theory, interaction, and objectivity are the main features of scientific method. In conclusion, teachers at the same time hold "traditional" and "modern" viewpoints on the nature of science. But, after after exposure to the research program, they were able to expand their nature of science webs to include more cultural aspects.

In brief, the main conclusions of the NOS research program were: (1) teachers have many difficulties in contradicting the neo-Kantian and positivistic dogma of the "axiological neutrality of science" and (2) through enhancing teachers' education with contemporary science, they would be enabled to meet its reflexive and ontological limits.

Level 2: School Science Text Books Studies

One of our research lines was to compare Greek and Cypriot teaching materials consisting of school science textbooks, workbooks, and teacher guidelines. Within this approach we studied current Greek[5] and Cypriot[6] nursery school teacher guide-lines, as well as the current Greek set of school science textbooks for the first grade.[7]

First stage: Following standards – Methodological notes

The textbook analysis consists of the following types:

1. Descriptive, concerning the target concepts, skills, and attitudes within the course "Studies of the Environment" for early grades.
2. Criterion analysis: We choose 11 criteria according to the benchmarks of the AAAS project 2061. Each criterion is characterized by a list of indicators. Each indicator is accompanied by a rating scheme. "Excellent," "satisfactory," and "poor" ratings are specified. Intermediate ratings of "very good" and "fair" are given in some cases. The total score of indicators shows how well the materials meet each criterion.

A brief evaluation shows the differences, similarities, and complementary potentialities of the studied materials for effective science education in early childhood.

Table 2. Indicative fundamental cross-thematic concepts in the compared textbooks.

		Greek nursery teachers guidelines	Cyprus nursery teachers guidelines	Greek teaching package for environmental Science-1st grade
Indicative Fundamental Cross-thematic Concepts	Human needs	X	X	X
	Individual-Group	X	X	X
	Space	X	X	X
	Communication	X	X	X
	Culture	X	X	X
	Time	X	X	X
	Environment	X	X	X
	Natural sciences	X	X	X
	Interaction	X	X	X
	System	X	X	X
	Similarity	X	X	X
	Differences	X	X	X
	Change	X	X	X

Selected Results. Stage 1. Descriptive Analysis

In the first phase of analysis, we recorded which concepts, skills, and attitudes are represented in the studied textbooks (Tables 2, 3, 4). Then, the criteria are described, along with a brief description of their fulfillment in each textbook (Table 5).

*Table 3. Comparison of the skills promoted in the
studied school textbooks.*

		Greek nursery teachers guidelines	Cyprus nursery teachers guidelines	Greek teaching package for environmental science-1st grade
Skills	Observation	X	X	X
	Classification	X	X	X
	Measurement	X	X	X
	Experiment	X	X	X
	Inferring	X	X	X
	Factor analysis	X	X	X
	Communication	X	X	X
	Modeling		X	X
	Prediction	X	X	X
	Hypothesis	X	X	
	Functional definitions		X	
	Data interpretation	X	X	

*Table 4. Comparison of attitudes in
studied school textbooks.*

		Greek nursery teachers guidelines	Cyprus nursery teachers guidelines	Greek teaching package for environmental science-1st grade
Attitudes	Environmental protection	X	X	X
	Humanitarianism	X	X	X
	Objectivity	X	X	X
	Volunteering	X	X	X

Table 5. Description of the criteria for analyzing school textbooks and a brief assessment of each criterion fulfilment.

		Greek nursery teachers guidelines	Cyprus nursery teachers guidelines	Greek teaching package for environmental science-1ˢᵗ grade
1.	The curriculum materials attempt to make their purposes explicit and meaningful to students.	yes	yes	yes
2.	Take account of student ideas.	yes	partly	Yes
3.	Provide multiple representations of the relevant phenomenon.	Yes	Yes	Yes
4.	Provoke practical work/field work.	Yes	Yes	Yes
5.	Develop scientific skills.	Yes	Yes	yes
6.	The materials lead to experimental processes.	Yes	Yes	yes
7.	The materials promote meta-cognitive and self-regulation processes.	yes	No	yes
8.	Promote creative reconstruction of data.	yes	no	yes
9.	Engage scientific views for the phenomenon under discussion.	yes	yes	yes
10.	Engage in authentic student questioning.	yes	no	yes
11.	Promote scientific literacy.	yes	yes	yes

Selected Results. Stage 2: Criterion Analysis.

In this stage some indicators are used to analyze each criterion, and each indicator follows a rating schema from "poor" to "excellent." In Table 6, some examples of the criterion analysis are presented.

Table 6.1 Criterion analysis on the clear sense of purpose and motivation for each activity.

Criterion 1[8]	*The course provides a clear sense of purpose and direction that is understandable and motivating for the children. This is done for each lesson or activity.*	**Rating Scheme** **Excellent:** *The material meets all indicators.* **Satisfactory:** *The material meets any three out of the five indicators.* **Poor:** *The material meets no more than one out of the five indicators.*		
		Greek nursery teachers guidelines	*Cyprus nursery teachers guidelines*	*Greek teaching package/1st grade*
Indicators	1. The material **conveys** or prompts teachers to convey **the purpose** of the activity to students. 2. The purpose is expressed in a way that is likely to be **comprehensible** to students. 3. The material encourages each student to **think about** the purpose of the activity. 4. The material conveys or prompts teachers to convey to students how the activity **relates to** the **unit purpose**. 5. The material engages students in thinking about **what they have learned so far** and **what they need to learn/do next** at appropriate points.	Satisfactory	Excellent	Satisfactory

Comments: Tables 6.1 and 6.2 show one part of the analysis for Criterion 1. This provides a sense of purpose. In Table 6.1 this sense of purpose is studied with respect to the lessons or the activity, and it has also been studied at the thematic unit level. In Table 6.2 the material is studied through the prism of providing a logical sequence of actions in relation to what extent the material conveys the rationale for this sequence. We can see that the Cyprus Nursery Teacher Guidelines excel. This can be interpreted as the more closed curriculum in Cyprus instead of the more open curricula in Greece. But we need a deeper study on how they convey more or less the rationale of each sequence of activities.

Table 6.2 Criterion analysis on the rationale of a sequence of actions.

	Justifying lesson/activity sequence. Does the material involve students in a logical or strategic sequence of lessons or actions (versus being just a collection of lessons or actions)?	***Rating Scheme*** ***Excellent:*** The material meets all indicators. ***Satisfactory:*** The material meets any three out of the five indicators. ***Poor:*** The material meets no more than one out of the five indicators.		
Criterion 1		*Greek Nursery Teachers Guidelines*	*Cyprus Nursery Teachers Guidelines*	*Greek Teaching Package/1st Grade*
Indicators	2. The material includes a **logical** or strategic sequence of actions. 2. The material conveys the **rationale** for this sequence.	*Excellent*	*Excellent*	*Satisfactory*

Criterion 2 refers to the extent that student textbooks take into account the children's ideas or previous sociocultural knowledge, and we present some examples from the school textbooks.

Example 1: Activity part that demonstrates the requirements of criterion 2: Taking into account the prior student's knowledge and experiences. Greek nursery school teacher guidelines.
Activity title: Investigating the pupil's prior knowledge and experience.
Topic: Water.
Activity description:
- The teacher motivates children to express their thoughts, knowledge, conceptions, ideas about the subject matter, what is the state of water.
- Teacher and children make a conceptual map, which can contribute to the teaching plan.
Questions to the children:
Who drinks water?
Where is the water?
What is the use of water?
Where do we put water?
What does water look like?
Why do we drink water?
How does water reach our homes?
What games can we play with water?
What happens when we play with water?
What are the sounds of water?
How do we make ice cubes?

Example 2: Activity part that demonstrates the requirements of criterion 2: Taking into account the prior student's knowledge and experiences. Cyprus nursery school teacher guidelines.

Activity title: Investigating the pupil's prior knowledge and experience.

Topic: Animals.

Activity description: Pre-lesson activities can be:

– Visit to a pet shop where children observe animals such as fish, rabbits, hamsters, turtles.
– Animal exhibition in the classroom, where children bring and demonstrate animals such as turtles, fish, parrots, ducklings, butterflies, chicks, bees, snails, ants, worms.

During the previous activities children have a chance to observe, to talk about their personal experiences, to draw, to sing, to imitate either the animal voices or their movement kinesthetically.

Table 7. Final comparison.

Greek Nursery Teachers Guidelines	Cyprus Nursery Teachers Guidelines	Greek Teaching package/1ˢᵗ grade
– Based on the schema environment-society-culture – Open Curriculum – Given didactic transformation – Doing emerged objectives – within the open curriculum frame. – Inter-disciplinarity and cross-disciplinarity. – Taking students ideas into account directly. – Emphasis on experience. – Literate transformation	– Based on the schema environment-society-culture – Close Curriculum – Given didactic transformation – Clear, understandable and motivating objectives. – Inter-disciplinarity. – Taking students ideas into account indirectly. – Emphasis on experience. – Literate transformation	– Based on the schema environment-society-culture – Medium Close Curriculum with some outcomes. – Given didactic transformation – Clear, understandable and motivating objectives. – Inter-disciplinarity and cross-disciplinarity. – Taking students ideas into account in case. – Emphasis on experience. – Literate transformation

Overall Comparison of the Materials

In Table 7 we can read a comparison of the trends and perspectives of the studied materials. A general comment is that the Greek materials seem to be more open than the Cypriot materials. This is because the Cypriot materials follow step-by-step instructions in order to promote scientific skills with accompanying references to the scientific domain and background of the activities. The Cypriot materials have a format that looks more scientific and more oriented to the teaching and learning of natural sciences. The Greek materials seem more interdisciplinary and more oriented to the children's citizenship, using natural sciences as a bridge. All textbooks fulfill the 11 evaluation criteria, but this study needs a deeper analysis that goes beyond the criteria and that is based on the implementation of those materials in real classrooms. Some case studies are described in the following paragraphs.

A Case Study on the Living Things: The Plants

This case study connects sociocultural theory with the aim of achieving scientific literacy. It tries to bridge the gap between academic approaches and citizenship approaches in teaching and learning natural sciences in the early grades. It focuses on meaning making when children interact with the physical-social-cultural world around them. It also tries to "open" natural sciences to society. Great importance is given to research based on Cultural-Historical Activity Theory (CHAT), which puts emphasis mostly on developing interactive networks of activity systems. In moving towards scientific literacy, CHAT offers a new field of research in Natural Science Education. The study presents some critical views of scientific literacy in modern science curricula and of the scientific content of natural science school textbooks. Moreover, a new mentality emerges in which science education is considered as part of society. This becomes important for the early grades, as it leads to a better understanding of scientific concepts and effectuates the aims of scientific literacy. The research is based on the school textbook "Studies of the Environment" for the first class in primary school and provides an example of achieving scientific literacy in two primary school classrooms.

The Research

During the academic/school year 2009–2010, especially during the second half of May, we carried out research in two public primary schools. We teach, observe, and record the unit "Flora and Fauna of Greece: Knowing the Plants of My Place" in the first grade. One of the schools was urban, located in the city of Ioannina, and the other was a rural school on the island of Corfu. The classroom teachers undertook all the teaching, while two researchers were observing, recording and videotaping the teaching praxis. The researchers worked only as observers and did not get involved in the teaching process. As a case study, the selected unit is strongly related to both science and environmental education. It is also one of the last units in the curriculum, and it is usually taught during the last months of the school year, so we were thus sure that the pupils/research subjects were able to read and write.

First school – Prefecture of Ioannina. The school class consisted of 25 pupils who were 6 to 7 years old. The teacher had 20 years of teaching experience. The experimental teaching lesson was conducted during the first teaching hour, lasted 45 minutes, and the pupils worked in groups.

The first lesson was about the plants of the pupils' homeland. It aimed to make the pupils capable of recognizing, naming, and describing the plants of their surroundings. The previous day the class had had an excursion to a nearby green park that was intended to arouse the pupils' interest in the subject matter. The pupils expressed their ideas and viewpoints about the plants in general and then the teacher motivated them to describe the characteristics of the plants they had referred to. In addition, the pupils were asked to classify the selected plants into two large categories: trees and flowers. Another classification activity was to match some plants with their names (Figure 4). Finally, the pupils were asked to recognize and discuss plant names that are first names of people (e.g. Daisy: a feminine name and daisy: a flower name). In this way they could correlate the teaching unit with themselves and their daily lives.

Figure 4. Activity from the school science textbook, p. 124 (Plakitsi. Kontogianni, Spyratou, & Manoli, 2006). The unit 1 title is "Which kind of plants live in my place?" In Activity 1 the children were being motivated by a question and a guideline: "Which kinds of plants do you recognize in the given pictures? – Connect each plant with its name."

2nd school – Prefecture of Corfu. The classroom consisted of 16 pupils from 6 to 7 years old. The teacher was a young person with 5 years of teaching experience. The experimental teaching lesson was carried out during the two first teaching periods, lasted 80 minutes, and the pupils worked in groups and individually as well.

This class did two lessons: The first was about plants in the pupils' homeland, and the second was about the parts of a plant. The second one aimed to make pupils capable of recognizing and naming the basic parts of a plant (roots, blastus, blooms) as well as to find similarities and differences in the plants' morphology. Our teaching equipment was one PC computer and a projector for presenting the worksheets of the school textbook.

The teacher and the pupils initially discussed the plants and explored the subject matter. The pupils expressed their viewpoints about the plants, and the teacher motivated them to describe some plant characteristics. They continued to identify some of the surrounding plants and to match them with their names. The pupils also discussed plant names that are used as human first names. The next activity was about plant classification based on where we find them. Twelve photos presented different kinds of plants and four categories of biotopes: field, garden, forest, water. Pupils did a classification collage activity in the school science workbook. The lesson continued with each pupil participating in the discussion about the parts of a plant and matching them with the appropriate picture in the school science textbook. Finally, pupils in pairs worked on leaf collections. The teacher randomly distributed 10 to 15 leaves to each pair and then asked them to classify them according to criteria such as width, length, color, and shape (Figure 5).

Figure 5. Students classify leaves using several criteria. (Photo from the authors' archives.)

Comments. The observation and analysis of the teaching led to some interesting comments on teaching science in the first grades. Our commentary focuses on the concept of scientific literacy and the way this is manifested in the current curriculum and school science textbooks in the context of modern sociocultural theory.

Below, we present a triangle analysis of an activity from the unit "Which Plants Live in My Place?" from the school science textbook, *Studies of the Environment for the First Grade*. The basic aim is to conceptualize scientific concepts as well as to develop some basic scientific skills, such as observation, classification, and communication. Some environmental attitudes as well as scientific values underlie the learning community rules and the division of labor. In Figure 6 we present some mediating tools and the way they influence activity development.

Unit 1: "Which Kind of Plants Live in My Place?" In the activity system presented in Figure 6 are the subjects of the activity, the rules, the communities, the knowledge content matter, the tools, and the division of labor. The school science textbook poses topics for discussion and problems, and in this way facilitates the internalization of knowledge, skills, and attitudes and leads towards scientific literacy. This mental function is directly supported by the rules.

Activity 1: "Which kinds of plants do you recognize in the given pictures? – Draw a line from the plant to a card with its name."

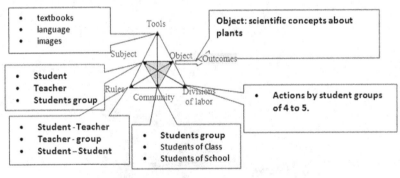

Outcomes:
- Scientific literacy
- Attitudes and values
- Scientific skills

Figure 6. Activity triangle analysis of the unit "Which Kind of Plants Live in My Place?"

The tools of this activity (school science textbook, communicative language, pictures with plants) work as mediators between the subjects (pupils and students) and the object (scientific concepts).

Furthermore, the learning community (school classroom) poses rules for inter-actions between students and the teacher as well as for the interactions that will develop (student-student, student-teacher, group-group, group-teacher, teacher-group). Students are divided into groups of four or five, and the work is distributed among them. The division of labor advances group effectiveness and increases the students' flexibility. The outcomes refer to the attitudes and values that develop and scientific literacy.

Discussion

It becomes obvious that the scope of teaching science is not just the learning of a specific corpus of scientific knowledge, but it is also the achievement of a wider scope for citizens of the 21st century, and that is scientific literacy. A curriculum that enhances scientific literacy must take into account the following:
- Its content must be based on the previous knowledge and experiences of the children.
- The knowledge and learning targets must be realistic and flexible so they can be modified according to class needs.
- The skills must be developed using many different topics related to the socio-cultural classroom environment and must refer to the nature of science.

Overall, from the case study combined with textbook criterion analysis, one can see that we started from a more positivistic critical analysis of the school science textbooks based on some criteria and standards and that we then shifted our focus towards a more cultural historical perspective. We implemented some units of the school science textbooks in two public schools, we videotaped the teaching, and observed the teaching/learning procedure through the prism of CHAT. It was obvious that even the school science textbook was open for societal/cultural/ environmental actions and that only a few of that these actions occurred at the school in Ioannina. From the teacher's interview we conclude that this happened because of the teacher's views of the nature of science, which were more traditional and positivistic. In contrast, in the school in Corfu, a more fruitful learning environment has been created. Many interactions occurred because of the teacher's views concerning the nature of science, which were more critical, cultural, and societal. These views had been reproduced and represented in the teachers' teaching practices. As researchers, it was easy for us to observe these differences.

In science education, learning concepts, teaching perspectives, and teacher and student identities form the whole pedagogical context. And these phenomena of interest are difficult to identify as cultural and historical in the field in which they appear. We test teachers' views on the nature of science by observing them as they teach and motivate students to learn about plants. Our observations showed that teacher training is crucial in order to achieve any shift towards a more cultural-historical perspective.

Level 3: A Leap Forward: Thinking of Learning and Teaching Phenomena as Cultural and Historical in the Field in Which They Appear

In order to think of learning and teaching phenomena as cultural and historical in the field in which they appear, we carried out research on dialogues among primary students concerning major environmental topics relevant to their local environment.

The main research question concerns the development and improvement of students' arguments in regard to terms such as verbal interaction, claims, warrants, data, and justifications. The analytic tools of inquiry that we used were (1) Kumpulainen (1999), (2) Toulmin (1958), and (3) Walton (1996). Students discussed some topics and prioritized them based on their interest in them. These topics were the school classroom environment, the school neighborhood, and Lake Pambotida in Ioannina, where the school is located. The research methodology followed the steps of action research (analyze, plan, act, observe, reflect) (Cohen, Manion, & Morrison, 2000). The research subjects were 100 students in the fourth and sixth grades of four public schools in the city of Ioannina, four teachers, and two researchers. The action research lasted three months. The results of research, even though limited, show a great potential for advancing students' argumentation in primary education. This research belongs to that which explores the modern dialogical shift in science education. We think that by studying argumentative operations when teaching and learning science, we can contribute to the current agenda for the dialogical shift in science. In the global literature, the dialogical turn is represented by some prominent scholars, including Bakhtin (1981), Bruner (1996), Harre and Gillet (1994), Lemke (2001), and Wells (2000). Furthermore, studies indicate that inside this particular frame of inquiry the training process can be studied as a social-cultural system of activities (Engeström, 1999). More specifically, learning in natural sciences is considered to be an adaptation and/or transformation via participation in progressively evolving discourses and practices (Roggof, 2003). An authentic learning environment to enhance this scope has two main characteristics: (1) to be a cooperative investigation and (2) to be dialogically oriented.

Planning. As mentioned before, the methodology followed the stages of action research (Cohen, Manion, and Morrison, 2000). Particularly, we created a working team consisting of the four teachers from four classes who participated in the research and two science education researchers. Teaching and learning settings were as follows: The students (1) construct their own group knowledge after discussion, debate, and consensus; (2) continuously interact with each other in their group of four or five members, with the other groups, with the teacher, and in the large classroom group while sitting in a circle and reflecting on what they have done; (3) participate in dialogues on local environmental issues, which are their own issues, since the schools are near Lake Ioannina; and (4) construct arguments for the environmental dilemmas in which they are involved because the lake strongly affects their lives.

Another important point of research planning was the collaboration between teachers and researchers. In this way we anticipated that a kind of osmosis would take place between the research world and the practitioners' world. In our case, teachers acted as researchers in collecting and analyzing data. Our research aims to contribute another view of environmental activism by enhancing the communicative language of environmental dilemmas through the enrichment of argumentative operations. In our approach we apply the following steps (Cohen, Manion, & Morrison, 2000):

- First stage–identification: Recognition, preliminary evaluation, and construction of the real and authentic local environmental problem
- Second stage: Preliminary discussion, construction of questions, emergence of a reforming perspective
- Third stage: Review of literature and further discussion
- Fourth stage: Modification of an identified area, redefinition of the environmental dilemma
- Fifth stage: Selection of research procedures, construction of aims and objectives
- Sixth stage: Evaluation procedures
- Seventh stage: Implementation of the project
- Eighth stage: Interpretation of data and conclusions

Objectives of research were:

- To study the verbal interactions between the students with respect to its operations
- To study the arguments that students use when they discuss the environmental problem and look for its solutions
- To study the changes in the argumentative operations and the verbal interaction after students/teachers involvement in action research processes

The inquiring project investigates the potential of environmental science education as an authentic learning environment and contributes to the advancement of verbal interactions and argumentative operations. The choice of environmental issues/problems/dilemmas was negotiated among the students and emerged through the following criteria (Roth, 1995, 2001):

The issue comes from the students' nearby physical environment, so each student can relate to it.

The neighbors' environment is approached spirally: class-school-school neighbourhood – Lake Pambotida of Ioannina – as a part of the school neighborhood.

Students discussed various environmental issues, such as recycling, lack of places of amusement, and traffic safety when coming to and leaving school. The first topic of discussion was selected when the mayor informed the school community about his willingness to build a high school or a gym near the primary school building. The dilemma "high school or gym" became the daily topic for discussion among the students. The most interesting topic for the students was the lake, since it related to their daily life, and a new dilemma concerning the uses of the lake emerged when it was learned that hydroplanes would connect the cities of Ioannina and Corfu by very short air routes. At this stage the class was separated into two big groups. Each group elected one coordinator, and both groups also agreed to choose a president for the debate. Each one of the groups selected one side of the debate, that is to say whether they would be for or against this operation of the hydroplanes. Students studied local newspapers delivered to them by the teachers and parents and then they interviewed the president of the Pamvotis Lake

Management Body (http://www.lakepamvotis.gr/). They also contacted many people and conducted a survey concerning the routing of hydroplanes. They interviewed people who would benefit from the lake's tourism activities and others who were professionals, lake fishermen, residents of the lake island, members of the nautical club of Ionnina, and the rowing team. After all this work, students prepared, organized, and participated in a debate entitled "Hydroplanes or Not on the Lake." The debate was audiotaped and then analyzed. Overall data was collected from interviews, photos, audiotaping, and videotaping. From this three-month project we present the analysis of the two dialogues that occurred in the classroom. As already mentioned, we used three analytical tools (Kumpulainen, 1999, for verbal interaction; Toulmin, 1958, for argument construction; and Walton, 1996, for hypothetic reasoning).

1. With Kumpulainen's analytical tool (1999), we approached the first objective of the research and studied the operations of verbal interaction. The processes of interaction are described in Appendix 1.

2. We then used Toulmin's analytics tool (1958). According to Toulmin, the basic elements of an argument in analytic reasoning are claims, data, excuses/explanations/well-founded reasons (warrants), and theoretical beliefs (backings). All arguments are supported by a probative base, that is, constituted by data, which relate to the claims, which depend on the warrants, which in their turn may depend on underlying theories or beliefs. In practice, the arguments depend on the field of application. Also in practice, the warrants that shape claims depend on the rules, the concepts, and the values of the field of application.

3. Finally, for the study of presumptive hypothetical reasoning, we used Walton's new dialectic[9] (1999). We supposed that some of the Walton schemata would be more suitable as tools for studying the argumentative operations.

The new dialectic is mainly concerned with the most common kinds of everyday arguments, and is based on presumptive reasoning rather than deductive or inductive logic. Presumptive reasoning takes the form of an inference in which the conclusion is a guess or presumption, accepted on a tentative basis, and subject to retraction as a commitment, should new information come in. The new dialectic shares many common features with the old dialectic of Plato and Aristotle, but is also different from it in other features. In the new dialectic, argumentation is analyzed and evaluated as used for some purpose in a type of dialogue underlying a conversational exchange. Each type of dialogue has its own standards of plausibility and rationality against which to measure the successful use of an argument. Thus the new dialectic has a relativistic aspect that makes it different from the classical positivistic philosophy. But it also has a structure with logical standards of evaluation of argument use, which makes it different from postmodern anti-rationalism. (Walton, 1999, p. 71).

For the needs of our analysis we modified Walton's 25 forms into four collapsed categories, which we analyzed.

Results – Comments. Some outputs from two dialogues are presented afterwards with the framework of their analysis.

1. *Analysis of two dialogues according to Kumpulainen (1999). The language functions in the verbal interaction are described in Appendix 1.*

Fragment 1 from Discourse 1 ("We discuss what we propose to build in the public area in front of our school: a HIGH SCHOOL or a GYM").

Nafsika: In our neighborhood we do not have a high school. So, next year we have to go far from our homes in order to go to high school. [I]

Nikos: Nor do we have a gym, and in order to exercise we must go to the opposite edge of the city. [AR]

Julie: This interests only you and not the majority of us. [Jd, E]

Nikos: You are mistaken. Sports interest many people. [Jd, E]

Julie: Well. [Ja]

Giorgos: There are some open athletic places very close by. You should go there. [I]

Vangelis: In those places we cannot exercise in the winter. [AR]

Iosif: We can exercise during the entire year [AR]

Xenia: Parents feel insecure about their children because the high school is far away. [AF]

Fragment 2 from Discourse 2 ("Hydroplanes in the lake? YES or NO?").

Nikos: I read that the lake is threatened. The lake's ecosystem is in danger from the hydroplanes, and the floats of the hydroplanes agitate the already very shallow waters of the lake. [I/Ja]

Giorgos: The birds of the lake will leave because of the noise, and they will be in danger from the hydroplane's propeller. Also, if the planes go near to the shore, big waves will be raised and can cause destruction to the birds' nests. [AR/I/Ja]

Balentina: Right now the boats that provide transportation to the island are much noisier, and they raise bigger waves than the hydroplane. [AR/Jd/E]

Costas: Also, the hydroplanes only have a few seats and are small machines. They make less noise than the big airplanes that provide air transportation from Athens to Ioannina and vice versa. [AR/Ja/E]

Julie: I read in the local newspaper that floats of hydroplanes are dyed with toxic dyes that will increase the pollution of the lake. [I]

Marios: From an interview with the owner of the hydroplanes, we were informed that that those dyes are totally harmless and are friendly to the environment. [I]

Thanassis: we should have proof that the hydroplane owner speaks the truth. [AR]

A first result from our analysis is that each one of the proposals of discourse 2 includes more language functions than the proposals of discourse 1. Also, the functions AR/Ja or Jb/E are presented more systematically in discourse 2. In our opinion, even from the very small fragment given, it is explicit that the children tended to analyze and argue about their opinions, rather than express a simple claim, which is often determined by their sentiments.

2. *Argument analysis according to Toulmin (1958).*

Figure 7 represents a first quantitative analysis of the distribution of basic argumentative operations in both discourses (red refers to discourse fragment 1, and green is for discourse fragment 2). From left to right, the first pair of columns represents the percentage of arguments with only one argumentative operation, the second pair of columns represents the percentage of arguments with two argumentative operations, the third pair of columns represents the percentage of arguments with three argumentative operations, and the fourth pair of columns represents the percentage of arguments with four argumentative operations. We observe that in the first fragment with regard to the construction of a high school or a gym in the neighborhood of the school, the arguments were characterized (P=65%) by only one argumentative operation. Twenty-two percent of the arguments are characterized by two basic operations, and only 13% are characterized by three basic operations. In the second dialogue (with regard to the routing of hydroplanes in Lake Pambotis) the four categories of analysis also exist. Arguments with one basic operation are recorded with a percentage of 23%; arguments with two basic operations, up to 31%; and arguments with three basic operations, up to 38%. Finally, only in the second discourse was there recorded a rich category with the four basic argumentative operations (8%). On a qualitative level, the categories of the basic argumentative operations appear in Table 8 which follows.

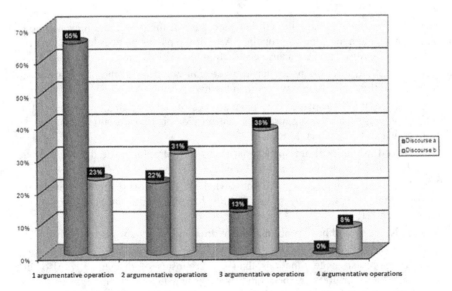

Figure 7. Percentages of the different types of arguments in discourses 1 and 2.

As we can observe when we compare the columns in Table 8, especially for the second discourse, a rich argumentation is recorded, with more explanations and justification of opinions, as well as underlying theories and beliefs. Also, while in the first dialogue only 5/17 language functions are recorded, with more than one basic argumentative operation, in the second dialogue 10/17 language functions are recorded.

Table 8. Discourse analysis based on Toulmin's argumentative operations.

	Discourse 1	Discourse 2
Nafsika	Data	Claim/Data
Nikos	Data	Claim/Data/warrants
Julie	Claim	Data
George Bagelis	Data	Claim/Data/warrants
Iosif	Claim	Claim/Data/warrants/backings
Xenia Costas	Claim	
Rania	Claim	Claim/backings
Bagelis B	Claim/Data	Claim/Data/warrants
Maria	Claim/warrants/backings	Claim/Data/warrants
Alexandros	Data	Claim
Balentina Stefania	Claim	
Marina	Warrants	
Thana Marios	Claim/Data	Claim/Data/warrants
	Claim	Claim/Data
	Claim/warrants	
	Claim/Data/warrants	Warrants/backings
	Claim	Claim

3. *Discourse analysis according to Walton (1996).*

The collapsed categories formed from the 25 schemes of argument proposed by Walton are:
Request for Information (it looks like this, it would be supposed ..., it could be ...).
Expert opinion (in the book he writes that ..., the scientist supports ...).
Inference: cause-effect, evidence-based inference.
Analogy.
Fragment analysis from discourse 1: "High school or gym:"
Analyzing discourse 1, we recorded arguments that belonged only in the first collapsed category. These arguments present a point of view, a personal opinion, a position. As examples, we report the following student arguments:

Julie: This interests only you and not us. [Personal opinion, position]
Nikos: You are mistaken. Sports interest many people. [Personal opinion-position]
Xenia: Parents feel insecurity for their children because the high school is far away from the neighborhood. [Personal opinion]

Fragment analysis from the discourse 2: "Hydroplanes in Lake Pambotis: yes or no?"

Analyzing students' arguments and applying the collapsed categories from the Walton schemes of argument, we recorded the following:

Expert Opinion = three arguments as follows:

Nikos: I read that the lake is threatened. The lake's ecosystem is in danger from the hydroplanes, and the floats of hydroplanes agitate the already very shallow waters of the lake.

Julie: I read in the local newspaper that floats of hydroplanes are dyed with toxic dyes that will increase the pollution of the lake.

Marios: From an interview with the owner of the hydroplanes, we were informed that that those dyes are totally harmless and are friendly to the environment.

Inference = three arguments as below: (In addition to the written phrases, we interviewed students and got some words such as "in conclusion," "hence," "consequently," and "therefore." In this way we classify the following phrases in the particular category of inference.)

Giorgos: The birds of the lake will leave because of the noise, and they will be in danger from the hydroplane's propeller. Also, if the planes go near the shore, big waves will be raised and can cause destruction to the birds' nests.

Balentina: Right now the boats that provide transportation to the island are much noisier and they raise bigger waves than the hydroplane.

Costas: Also, the hydroplanes have only a few seats and are small machines, and they make less noise than the big airplanes that provide air transportation from Athens to Ioannina and vice versa.

Request for information/evidences = one proposal:

Thanassis: We should have proof that the hydroplane owneris telling the truth.

Studying all of discourse 2, we recorded one argument that could be put in the fourth collapsed category of Analogy:

Rania: But what do you say now? Approximately 20,000 residents of the Ioannina basin aren't connected with the sewage system yet (!), and you talk about toxic colors from a hydroplane?

As for conclusions concerning this case study on argumentation analysis, and taking into consideration the overall discourse analysis, we reached the following conclusions: first we used Kumpulainen's analytical tool about verbal interactions, and we recorded many more interactions in discourse 2. According to our second analysis using Toulmin's analytical tool, we found that: (a) the arguments that contained both data and warrants operations increased in discourse 2; (b) in both fragments of the studied discourses there are only a few arguments with backings and (c) we observed an accumulation of data operations in the first sample. Finally, and applying the analysis according to Walton's schemes of argument, it seems that the students' arguments can be improved through better documentation of their personal opinions. This documentation is based on data, evidence, expert opinions, and analogical reasoning when students are involved in authentic environmental dilemmas in their local environment. However, it is difficult to use Walton's schemes of argument (25 categories of analysis) in educational research; and this issue requires further investigation.

In our opinion, to overcome the limits of the several research tools, we need a more fruitful context of analysis (or analytic category) through which we can analyze and/or design environmental science education activities. The nature of environmental dilemmas and the nature of environmental activities are full of interactions, support the dialogic turn in science education, and are embedded in the sociocultural theory frame of learning. We propose that a further investigation of the terms and the ways the interactions occur and function is needed.

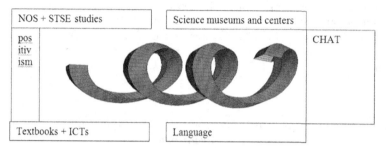

Figure 8. Mutiple helix representing the research evolution of the ATFISE group.

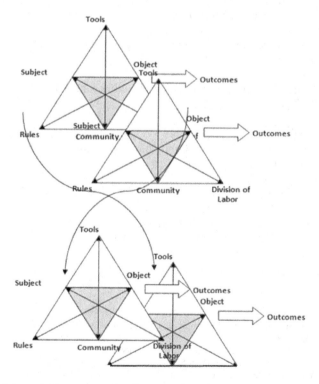

Figure 9. Joint activities representing the research evolution of the ATFISE group.

Overall. Our team gradually moved from a background in the positivistic aspects of science education to a more cultural-historical perspective. This evolution can be represented by a multiple spiral where different approaches led us to the development of a helix, which changes itself over time.

This evolution was not a linear process as might be implied by Figure 8, but rather a process of osmosis among some elements of some interactive systems of activities as in Figure 9.

The interactive systems of activities led us to seek a new analytic category within the framework of activity. This may be the concept of activism, which can highlight science education and overcome the danger of a self-referential circle of sociocultural analysis. This issue may become the subject matter of another forum.

APPENDIX I

A Part of the Kumpulainen Analytical Tool (1999) Concerning the Dimension of Language Functions during a Verbal Interaction

Language functions
Informative (I) Providing information
Reasoning (RE) Reasoning in language
Evaluative (EV) Evaluating work or action
Interrogative (Q) Asking questions
Responsive (A) Answering questions
Organizational (OR) Organizing or/and controlling behaviour Judgmental

Agrees (Ja) Expressing agreement or

Disagrees (Jd) disagreement

Argumentational (AR) Justifying information, opinions or actions
Compositional (CR) Creating text
Revision (RV) Revising text
Dictation (DI) Dictating
Reading aloud (RE) Reading text
Repetition (RP) Repeating spoken language
Experiential (E) Expressing personal experiences
Affectional (AF) Expressing feelings

NOTES

1. In Greece there is a national curriculum http://digitalschool.minedu.gov.gr/info/newps.php
2. IHPST: International History, Philosophy and Science Teaching Group.
   EARLI: European Association for Research on Learning and Instruction.
   ISCAR: International Society for Cultural and Activity Research.
   IOSTE: International Organization for Science and Technology Education.
   NAESA: Joint North American – European and South American Symposium on Science and Technology Education.

[3] The original table can be found in Bartholomew et al. (2004).

[4] Aikenhead's instrument is named Views on Science Technology Society (VOSTS).

[5] Dafermou, Ch., Koulouri, P., & Pasagianni, E. (2006). *Guidelines for the nursery teachers*. Athens: OEDB [in Greek].

[6] Constantinou, K.P., Feronimou, G., Kyriakidou, E., & Nikolaou, Ch. (2004). *Science in nursery school: Teacher guidelines*, 2nd ed. Nicosia: Ministry of Education and Culture of Cyprus [in Greek].

[7] Plakitsi, K., Kontogianni, A., Spyratou, Ei., & Manoli V. (2006). *Studies of the environment*. Student book, workbook and teacher guidelines. Athens: OEDB [in Greek]. This set is based on the Cross – Thematic curriculum Framework for Citizenship Education. Relevant sites: http://www.pi-schools.gr/download/programs/depps/english/22th.pdf, http://www.oedb.gr:8080/portal/portal/default/default.

[8] The criterion and the indicators are based on the benchmarks of the project 2061. http://www.project2061.org/publications/textbook/mgsci/report/crit-used.htm.

[9] http://www.dougwalton.ca/papers%20in%20pdf/99newdial.pdf.

REFERENCES

Bakhtin, M. M. (1981). *The dialogic imagination* (M. Holquist, Ed.). Austin, TX: University of Texas Press.

Bartholomew, H., Osborne, J., & Ratcliffe, M. (2004). Teaching students 'Ideas-about-science': Five dimensions of effective practice. *Science Education, 88*(5), 655–682. (10.1002/sce.10136).

Bruner, J. (1996). *The culture of education*. Cambridge, MA, London: Harvard University Press.

Cohen, L., Manion, L., & Morrison, K. (2000). *Research methods in education* (5th ed.). London: RoutledgeFalmer.

Engestrom, Y. (1999) Activity theory and individual and social transformation. In Y. Engestrom, R. Miettinen, & R. L. Punamaki (Eds.), *Perspectives on activity theory* (pp. 19–38). Cambridge, UK: Cambridge University Press.

Goulart, M. I. M., & Roth, W. M. (2010). Engaging young children in collective curriculum design. *Cultural Studies of Science Education, 5*(3), 533–562.

Harre, R., & Gillet, G. (1994). *The discursive mind*. London: Sage.

Kokkotas, V., & Plakitsi, K. (2005). *Time for education: Ontology, epistemology and discursiveness in teaching fundamental scientific topics*. Paper presented to 1st International Conference of International Society for Cultural and Activity Research (I.S.C.A.R.), Seville, Spain.

Kokkotas, P., & Vlachos, I., in collaboration with Hatzi, M., Hatzaroula, V., Plakitsi, K., Rizaki, A., & Verdis, A. (1997). The role of language in understanding the physical concepts. 2nd World Congress of Nonlinear Analysts. Athens, July 10–16. *Nonlinear Analysis, Theory, Methods & Application, 1997, 30*(4), 2113–2120.

Kuhn, D. (1993). Science as argument: Implications for teaching and learning scientific thinking. *Science Education, 77*, 319–337.

Kumpulainen, K., & Mutanen, M. (1999). The situated dynamics of peer group interaction: An introduction to an analytic framework. *Learning and Instruction, 9*, 449–474.

Lemke, J. L. (2001). Articulating communities: Sociocultural perspectives on science education. *Journal of Research on Science Teaching, 38*(3), 296–316.

Machamer, P. (1998). Philosophy of science: An overview for educators. *Science & Education, 7*, 1–11. [Also found in Bybee, R. W., et al. (Eds.). (1992). *Teaching about the history and nature of science and technology: Background papers*. Colorado Springs, CO: BSCS.]

McComas, W. F., Clough, M. P., & Almazroa, H. (1998). The role and character of the nature of science in science education. *Science & Education, 7*(6), 511–532.

Piliouras, P., Plakitsi, K., & Kokkotas, P. (2007). *Sofia doesn't speak during team work. Using discourse analysis as a tool for the transformation of peer group interactions in an elementary multicultural science classroom*. Paper presented to 12th Biennial Conference for Research on Learning and Instruction, EARLI 2007, Budapest, Hungary.

Plakitsi, K. (2008). *Science education in early childhood: Trends and perspectives*. Athens: Patakis [in Greek].

Plakitsi, K. (2010). Collective curriculum design as a tool for rethinking scientific literacy. *Cultural Studies of Science Education*, 5(3), 577–590.

Plakitsi, K. (Ed.). (in press). *Sociocognitive and sociocultural approches in early childhood*. Preface: Wolff-Michael Roth (Authors: Wolff-Michael Roth, Reggio Emilia team leaded by Olmes Bisi, Agustín Adúriz-Bravo, and all the university professors of Science Education in the Greek Early Childhood Departments). Athens: Patakis [in English and in Greek].

Plakitsi, K., & Kokkotas, V. (2004). Review on the epistemological approaches of nature of science in modern educational systems. Study on the epistemological approaches through schoolteachers' beliefs and practises. *Book of Proceedings*. Learning & Teaching in Knowledge Society, Greece: University of Athens.

Plakitsi, K., & Kokkotas, V. (2005) Reflective, informal and non-linear aspects of argumentation in school practice. In *Proceedings Book* of the 4th I.O.S.T.E. Congress, Greece, 2004, Ed. E.DI.F.E.

Plakitsi, K., & Kokkotas, V. (2006). *Enhancing teachers' education through interpretive-philosophical meditation about the Nature of Science: The MaPrOject*. Paper presented to the Joint North American – European and South American (N.A.E.S.A.) Symposium *Science and Technology Literacy in the 21st Century*, May 31–June 4, 2006, University of Cyprus. Proceedings, Vol. 1, pp. 200–211.

Roggof, B. (2003). *The cultural nature of human development*. Oxford: University Press.

Roth, W. (1995). *Authentic school science: Knowing and learning in open-inquiry science laboratories*. Dordrecht, Netherlands: Kluwer Academic Publishing.

Roth, W. M. (2001). Situating cognition. *Journal of the Learning Sciences*, 10(1), 27–61.

Toulmin, S. (1958). *The uses of argument*. Cambridge: Cambridge University Press.

Walton, D. (1996). *Argumentation schemes for presumptive reasoning*. Mahwah, NJ: Lawrence Erlbaum Associates.

Walton, D. (1999). The new dialectic. *Protosociology*, 13, 70–91.

ATFISE Group
(K. Plakitsi, E. Kolokouri, E. Nanni, E. Stamoulis, & X. Theodoraki)
School of Education
University of Ioannina
Greece

EFTHYMIS STAMOULIS & KATERINA PLAKITSI

6. ACTIVITY THEORY, HISTORY AND PHILOSOPHY OF SCIENCE, AND ICT TECHNOLOGIES IN SCIENCE TEACHING APPLICATIONS

INTRODUCTION

This paper is part of a study which focused on the restructuring of scientific literacy (Roth & Lee, 2004), in the sense that it proposes a new methodological tool for designing the teaching of scientific concepts to young children. This new approach comes under the umbrella of activity theory. However, if activity theory is employed mainly in the analysis of activities (Engeström, 1987), in the present study it is employed in an attempt to design the activities of students.

In the present study, we apply the expansive cycle which Engeström (1999b) suggests and which he accepts as the acceptable practice of an action, and step by step we are led – through the resolution of contradictions – to the application of new practices. These new practices help us to design activities in the teaching of Science to pupils of primary education. In implementing these we use elements derived from the History of Science and teaching practices which are suggested in the literature and are relevant to their introduction in the teaching of science.

All activities, implemented by the pupils with the help of ICT, do not take place in the traditional environment of a class. Much research on this subject has been undertaken internationally. It shows that the incorporation of elements from the History of Science and the use of ICT in the teaching of science to young pupils stimulate the interest of the children and enhance the understanding of the concepts of science. In this study, following an exposition of activity theory, as propounded by its advocates, we describe the contribution of the elements of history and philosophy of sciences in education, as well as the ways which are suggested for their introduction. Furthermore, the necessity which has arisen at present for the use of new technologies in education is also analysed.

The aim of this study is to offer teachers a new methodological tool for the analysis of the activities of the pupils as well as the planning and designing of such activities with the intention of upgrading the quality of the teaching of science.

As a case study for the design of activities we chose the concept of electromagnetism. We followed the historical development of the idea from ancient times – when electric and magnetic phenomena were clearly distinguished, until their unification by Ørsted and Faraday. For the collection of the data of the research a variety of methodological tools was used (videos, works by other students) and their analysis took place under the main tenets of activity theory.

K. Plakitsi (ed.), Activity Theory in Formal and Informal Science Education, 111–157.

THEORETICAL FRAMEWORK

The Framework of Activity Theory

Cultural-historical psychology, as formulated by scholars representing many national traditions begins from the assumption that there is an intimate connection between the specific environment human beings inhabit and the fundamental, distinguishing qualities of human psychological processes. Activity theory (AT) is a social psychological theory about the development and the dynamics of collective human activity.

Activity theory provides a context of human activity and proposes a set of practices which link individual to social activity (Engeström, 1999a; Nardi, 1996).

The fundamental structural unit of the theory is human activity taking place in the smallest possible context necessary to understand it and includes the subject – which is a person or a group (e.g. a group could consist of a student or a group of pupils or students and teachers) performing an activity on an object – as well as the dynamic relationships which develop among them and the socio-cultural rules governing the activity (Barab et al., 2003; Cole, 1999; Engeström, 1999b; Kaptelinin et al., 1999).

Davydov supports the view that an activity depicts a specific aspect of social coexistence of people and this activity is a deliberate change of physical and social reality (Davydov, 1999). Wolff-Michael Roth (2009) calls for the inclusion of sensuous aspects of work in the unit of analysis. He names emotions, identity, and the ethico-moral dimensions of action, such as salient sensuous aspects. Roth suggests that sensuous aspects may be approached by focusing on actions together with their effects.

Activity theory provides all the necessary tools for a theoretical and methodological approach in the design and analysis of educational activities, which include much more than a tool that mediates between the subject and the objectives of teaching (object) (Barab et al., 2003). As we argued in our earlier paper (Stamoulis & Kokkotas, 2006) it is not sufficient to design a tool for an activity; it must be a tool that will be in the service of social interaction. We should not be designing tools but plans for social participation, especially with the use of computers, where the emphasis is transferred from human-computer interaction to human-human interaction. These tools of improved subject-object interaction involve some core issues, such as agent structure, communication, group awareness, consistency maintenance, and collaborative tasks.

Activity theory has applications in different fields of educational psychology (Koschmann, 1996), in human-computer interactions (Kuutti, 1996; Nardi 1996) and in the design and analysis of educational activities (Barab et al., 1999; Johanssen & Rohrer-Murphy, 1999; Kaptelinin et al., 1999). The model description of each activity is suitable for mapping the observed actions of subjects in ethnographic studies (Barab et al., 2004; Engeström, 1999b; Mwanza, 2000).

Engeström (1999b) suggests that we may distinguish between three theoretical generations in the evolution of cultural-historical activity theory.

The first generation, centered around Vygotsky, created the idea of mediation. This idea was crystallized in Vygotsky's (1978, p. 40) famous triangular model of 'a complex, mediated act' which is commonly expressed as the triad of subject, object, and mediating artefact. The insertion of cultural artefacts into human actions was revolutionary in that the basic unit of analysis now overcame the split between the Cartesian individual and the untouchable societal structure. The individual could no longer be understood without his or her cultural means; and the society could no longer be understood without the agency of individuals who use and produce artefacts. This meant that objects ceased to be just raw material for the formation of the subject as they were for Piaget. Objects became cultural entities and the object-orientedness of action became the key to understanding human psyche. [...] The concept of activity took the paradigm a major step forward in that it turned the focus on complex interrelations between the individual subject and his or her community. [...] Ever since Vygotsky's foundational work, the cultural-historical approach has been very much a discourse of vertical development toward 'higher psychological functions.' Luria's (1976) cross-cultural research remained an isolated attempt. Michael Cole (1988; see also Griffin & Cole, 1984) was one of the first to clearly point out the deep-seated insensitivity of second generation activity theory toward cultural diversity. When activity theory went international, questions of diversity and dialogue between different traditions or perspectives became increasingly serious challenges. It is these challenges that the third generation of activity theory must deal with. The third generation of activity theory needs to develop conceptual tools to understand dialogue, multiple perspectives and voices, and networks of interacting activity systems. In this mode of research, the basic model is expanded to include minimally two interacting activity systems.[1]

It should be noted that ten years later Engeström suggests that this theory should be expanded to address the various objections that have been raised against its major tenets and that it should be permitted to find its own. It is in this context that we are suggesting in this work that this theory should be developed in the direction of planning and designing of activities.

Learning, knowledge, and expertise – according to the activity theory – are effectively distributed through the participation of people in the community. This theory questions the notion of authority and specialization residing within an individual as opposed to such notions residing in a wider social context. So a new system for studying the intra-individual activity, the interpersonal level and the wider community is proposed (Engeström, 1999b; Leontiev, 1979).

Learning is considered to be a process of social interaction at micro, medium and macro levels. Co-operating with others, a person develops skills and abilities. Central roles in the process are (a) collaborating and (b) language as a tool which shapes the identity of the individual (Leontiev, 1979; Nardi, 1996; Vygotsky, 1978). Within this framework, the teaching approach is characterised by complex group work and peer tutoring, and the course of development is characterised by increasing levels of awareness, control, and consciousness of higher intellectual functions such as problem solving and reasoning (Wertsch, 1985).

How does the theory of activity support such teaching strategies? What methodology should be followed for planning activities? What actions are necessary to help students internalise the meaning/object of the activity? What is the use of tools? In this study we try to define a framework which may support activity theory for teaching science concepts to primary students. Through the utilisation of data from the History and Philosophy of Science and with the help of computers, we developed a specific application which aims at facilitating the students in the construction and comprehension of the abstract concepts of electromagnetism through an ICT-based instructional package.

According to Engeström the basic concept of activity theory is that learning is a human activity which is socially embedded and mediated by tools. The main idea is that some mental activities such as thinking arise from practical activity, and thus the unit of analysis should include the individual and the social-cultural context in which that activity takes place. The tools mediate the process between subject and object, the rules mediate the processes between the subject and the community, and part of the division of labour mediates the procedures between the community and the object (Figure 1). In other words, tools are used by subjects to achieve an objective, rules are set among the subjects and other members of the community in order to achieve the objective and community members need a division of labour in order to achieve the objective (Engeström, 1987, 1999b, 2001). The objective, in our case, is the construction of an object (e.g., a battery) by the students with the help of computers.

"Object" is a subtle concept: it has material form (i.e., it is a real, material object that is being worked on), but it also embodies the "idea" of the activity, i.e., it has an "ideal" form, that is, the motive, collective purpose, or envisioned outcome of activity (Roth & Lee, 2007). Activity theory is a theory of object-driven activity. Objects are concerns; they are generators and foci of attention, motivation, effort, and meaning. Through their activities, people constantly change and create new objects. Often, the new objects are not intentional products of a single activity but unintended consequences of multiple activities (Engeström, 2009).

Engeström's model shows that the main purpose of an object directed activity in education is the acquisition of knowledge and skills by students. The systemic model of Engeström focuses on three mutual inter-relationships involved in each activity: the relationship between subject and object, the relationship between subject and community, and the relationship between the community and the object (Bottino et al., 1999).

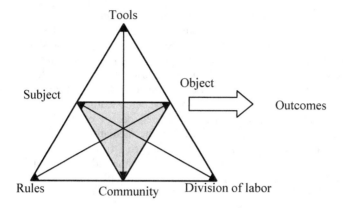

Figure 1. Activity system model (Engeström, 1987).

Mediation is the key to the activity theory. It is the mechanism through which external socio-cultural activities are transformed into internal mental functioning. The source of mediation may be a tool kit (for example, a rope around somebody's hand as a reminder), a system of symbols (language), or behaviour of another human social interaction. Mediators, in the form of objects, symbols, and people transform natural, spontaneous reactions into higher mental processes, as shown in the the the diagram above. In the case of science, this intervention may take the form of a handbook, an experimental set-up, a simulation, a material from the history of science or any help from the teacher (Basharina, 2007).

The principle of mediation derives from the work of the Soviet school of Vygotsky, Leontiev, Luria and other works which try to prove the validity of their theory. Mediation activities of the cultural tools (artefacts) and signs are not just psychological concepts, but the key link of the individual with the cultural environment and society.

Vygotsky contemplates the tool of mediation as an auxiliary stimulus which "promotes psychological operation to higher and qualitatively new forms, and allows people, with the help of external stimuli, *to control their behaviour from the outside.* The use of signs then leads people to a specific structure of behaviour that goes beyond organic growth, which creates new forms of culturally mediated psychological processes" (Vygotsky, 1978, p. 78, italics in original).

The idea that people can control their behaviour "from the outside," by creating and using cultural tools of mediation rather than through "internal" factors based on mental functions, could be considered as an optimistic perspective on human self-determination but also as a challenge, a call for study and research on cultural tools as integral components of human functioning (Engeström, 1999b).

Lee and Smagorinsky (2000, p. 3) summarise key aspects of Vygotsky's activity theory as follows:

    i. Learning, on an inter-psychological level, often involves mentoring provided by culturally more knowledgeable persons, usually elders, who engage in activity with less experienced or knowledgeable persons in a process known as "scaffolding" (Bruner, 1975). Knowledge is not simply handed down from one to the other, however. Meaning is Knowledge constructed through joint activity rather than being transmitted from teacher to learner.

    ii. Concepts or meanings are constructed out of content knowledge, strategies, and technologies, that is, the mediating tools or artefacts that are drawn on in the act of meaning construction – historically and culturally; [...] that is, individuals are connected to cultural history and its manifestation in everyday life. People, tools, and cultural constructions of tool use are thus inseparable [...].

    iii. The capacity to learn is not delimited. Rather, the potential for learning is an ever-shifting range of possibilities dependent on what the cultural novice already knows, the nature of the problem to be solved or the task to be learned, the activity structures in which learning takes place, and the quality of this person's interaction with others. [...] Vygotsky (1978) argued that because learning takes place in this zone of proximal development (ZPD), teaching should not extend beyond what a student cannot do without assistance, and not beyond the links to what the student already knows. (Lee & Smagorinsky, 2000, p. 3)

So, the ability of a pupil partaking in the learning process is not delimited. The potential for learning is a continuous process which depends on the dynamics of a society, which is "creating ever new forms of joint activities and the cognitive process takes place both between individuals and within themselves" (Cole & Engeström, 1993, p. 43). On this basis, Marx Wartofsky (1979, in Engeström 1999a), found that cultural mediation tools promote cultural changes, in a way analogous to the role of genes in biological evolution. Activity theory has the conceptual and methodological potential to guide studies which facilitates some control of the cultural tools of humans, and thus of their future.

Wartofsky's work in historical epistemology has exerted a decisive influence on the course of cultural-historical activity theory, as Engeström himself admits (1987). Wartofsky (1979) distinguishes three kinds of artefacts that can function as representations. First are material tools and the social practices in which they are employed; these are primary artefacts in the sense that they are directly involved in the transformation of the environment for the production and reproduction of the means of existence. The first such artefacts were simple tools (knives, spears, and pots); today, they include aircraft, computers, and automatic banking machines. Such artefacts are not created with the purpose of representing, but they can be so used, particularly to represent the activities in which they are typically involved. The second category consists of those that are created with the purpose of preserving the tools and practices by means of which primary activities are organised, and their motives, goals, and knowledgeable skills are passed on to new participants. These secondary artefacts are symbolic representations of the primary activities that are used to plan, manage, and evaluate. Face-to-face mimetic acts would have been the earliest form of secondary artefacts; nowadays they may be in one of a variety of semiotic modes or even in a combination (Wells, 2000).

> Such representations, then, are reflexive embodiments of forms of action or praxis, in the sense that they are symbolic externalizations or objectifications of such modes of action – 'reflections' of them, according to some convention, and therefore understood as images of such forms of action – or, if you like, pictures or models of them. (Wartofsky, 1979, p. 201)

In Wartofsky's classification there is also a third level of artefacts, abstracted from their direct representational function, called tertiary artefacts. Related to the division between "psychological" and "technical" tools made by Vygotsky, language cuts across all three different levels. These tertiary artefacts are the imaginative, integrative representational structures (myths, works of art, as well as theories and models) with which humans attempt to understand the world and their existence in it (Wells, 2000). However, there are interactions, transitions, and transformations between the physical and conceptual dimensions of objects rather than a clear and identifiable separation (Hakkarainen, 2004). With this simple taxonomy of artefacts based on their relation to production Wartofsky offers an alternative approach to this separation.

For Vygotsky, a child's development is above all a matter of becoming a fully functioning (in my terms, autonomous) member of a particular human culture. Social interaction plays a decisive role in this process, providing structures that are gradually internalized as cognitive capacities:

> Any function in the child's cultural development appears twice, or on two planes. First it appears on the social plane, and then on the psychological plane. First it appears between people as an inter-psychological category, and then within the child as an intrapsychological category. [...] Social relations or relations among people genetically underlie all higher functions and their relationships. (Vygotsky 1978, p. 104)

How we learn to think, in other words, is determined by the interactive structures in which our early experience is embedded. This argument helps to explain the coexistence in humanity of biological unity and cultural diversity. It also implies that to be optimal all human learning may require a social dimension, and it clearly relates our psychological autonomy to the interdependent processes of social interaction.

Vygotsky also believes that children in their early stages of development perform an activity mediated by a physical or symbolic tool, in co-operation with an adult, to get successful results. The external tool/item needed by children of school age is transformed into an internal point, which is revoked and subsequently used by the person in discussions or in the exercise of other activities. People are involved in many activity systems and, in their own microcosm, the outcomes of their performance are incorporated into the wider cultural context. A person is thus affected by other activity systems in which he or she participates. These influences are horizontal, occurring in small and large communities and perpendicular, incorporating the history, culture and relationships formed in human production activities (Basharina, 2007).

Principles of activity theory. We can distinguish five basic principles of activity theory (Engeström, 2001; Rizzo, 2003) that have been used in recent studies to analyze the success of the mediation of educational materials developed in computer environments and learning situations in which subjects-students with practical cooperative activities are performed to learn. These principles are:

1. The first principle is that a collective, artefact-mediated and object-oriented activity system, seen in its network relations to other activity systems, is taken as the prime unit of analysis (Engeström, 2001, p. 136).

The smallest structural unit of the theory is the activity – a collective system which is mediated by cultural tools and is object-orientated – which is seen in a network of relationships with other activity systems and is the smallest framework for understanding human actions. The activity is treated as the primary unit of analysis. Individual targets, actions of groups, automatic features that can be second-order units of analysis are fully understood only when interpreted within all dimensions of activity.

2. The second principle is the multi-voicedness of activity systems. An activity system is always a community of multiple points of view, traditions, and interests.

A system activity is by definition a multilevel (multi-voiced) formation. It is the orchestrator of different views and approaches of various participants (Engeström, 1999a). The activities are long-term formations whose objects are transformed into results through a process which typically consists of major steps and phases (Jonassen & Roher-Murphy, 1999). The upper level of a "collective activity" is guided to a target/object, the middle level of "actions" of an individual or a group is guided by a conscious purpose, and the last stage of the automatic features is guided by the conditions and tools of the above actions (Engeström, 1987, 1999a).

The activities are carried out as individual and collaborative actions and a series of actions or networks connected to each other via the same overall object and motive. Participation in an activity of this type means making conscious actions that have (and are directed by) a direct and specific target (goal). Every action is first designed by an individual consciousness in a standard way through the use of a mental representation (Engeström et al., 1999; Vygotsky, 1978). For example, an experimental procedure can be represented in Figure 2 (Leontiev, 1978).

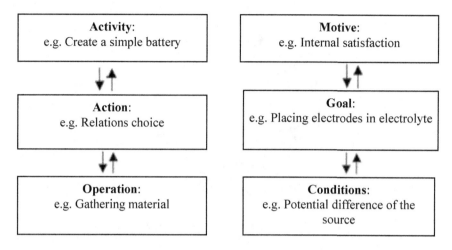

*Figure 2. Diagram representation of the hierarchical structure
of the experimental procedure.*

3. The third principle is historicity.

A key objective of historical analysis is frequency. Longer periods of history have the same characteristics as shorter periods or events, whose analysis gives us new levels of repetitive or cyclical time structures. In this respect, Engeström (1999b) supported and introduced the term expansive cycle.

If we form a bow to depict the repeated cycles or expanse, it is important to note that the time of the activity is qualitatively different from the time of action. The time for action is essentially linear and expects a finite end. The timing of the activity is repetitive and circular.

The expansionary cycle identified two key processes that occur constantly at each level of human activity: internalization and externalization. Internalization is associated with the reproduction of culture; externalization is associated with the creation of new artefacts which make a transformation of the culture possible (Figure 3).

The expansionary cycle of an activity system begins with an almost exclusive emphasis on internalization, socialization, and training students to become capable members of the activity as conducted. Creative externalization occurs first as specific individual innovations. Since the divisions and contradictions of the activity become more demanding, internalization increasingly takes the form of critical mass, and externalization seeks only strategic solutions. Externalization reaches its peak when a new model is designed for the activity and it is applied. When the new model is stabilized, the internalization of innate instruments is still the dominant form of learning and development.

Figure 3. The expansive cycle (Engeström 1999a, p. 34).

These two processes are inextricably interwoven. "It's no longer true to say that people create this [society]. Probably we should mention that people reproduce and transform" (Engeström & Miettinen, 1999, p. 10). A historical perspective would include previous cycles of the activity system.

4. The fourth principle is the central role of contradictions as sources of change and development.

The idea of internal contradictions as the driving force of change and development in activity systems is gaining ground as a basic principle of empirical research (Engeström, 1987). These activities are not isolated units; they are affected by other activities. The external influence could change some elements of activities causing contradictions (problems, ruptures, interruptions). Activity theory uses the term "contradictions" to denote the non-adjustment of data between each of the different activities, or between different development phases of a single activity (Kutti, 1996). In activity theory, contradictions are considered as a source of growth (Bottino et al., 1999) and are the driving force of change (Basharina, 2007).

The same objective is achieved in different ways using the same tools available in different sociocultural contexts. When people from different cultural areas are involved in the exercise of an activity to achieve a goal, the use of tools can interpret a heterogeneous set of communicative practices with different rules. However, the division of labour is not the same, but differs in both the communication style and the personal style of the participants (Thorne, 2003, p. 41). For example, communication tools, Internet/cultural tools are different objects for different communities, leading to diversification effects, communication processes, building relationships, and language development.

5. The fifth principle proclaims the possibility of expansive transformations in activity systems. Activity systems move through relatively long cycles of qualitative transformations.

Engeström (1999b), by defining the concept of "expansive cycle," begins with the accepted practice of an action or activity and leads us gradually into a collective movement or state. This movement is achieved through specific and targeted actions. All actions undertaken in this way form an expansive cycle during which tensions created in a system of activities are dealt with successfully.

This approach is similar to the approach of the theory of networks (Latour, 1988; Bigum, 2000), in which learning is a heterogeneous network between people. Engeström (2001) stresses that there is an interactive approach between activity theory and the theory of networks, but the main tenet (in the theory of networks) leads to the degradation of people with no identifiable internal systemic properties and without contradictions. Activity theory differs from the theory of networks, as it provides a comprehensive instructional framework based on learning and professional development of people when it takes place in a structured environment, such as the "school," and is mediated by new technologies.

This is particularly important because new technologies have been introduced so far in an organized learning environment of the school based on traditional teaching practices. This has led, at best, to a marginal effect by new technologies in everyday workplace learning in the classroom. The fundamental assumption of this approach is based on the pattern of transmission of knowledge from teacher to student. Knowledge and skills are treated as a finite and well-specified object that can be handled well by a "teacher" (be it human or computer) and totally disconnected from any social and cultural context in which the underlying act exists. It introduced the activity theory and expansive learning is what violates the model described above (Engeström, 2001). Teachers and the organized learning system of schools meet the challenge posed by new technologies, and all learn something that is stable and clearly defined but may change gradually over time. In other words, as teachers, during our participation in the implementation of activities, we must learn new forms of activity not previously shown. They are learned as they occur. There is no "competent" teacher who has knowledge.

> A full cycle of expansive transformation may be understood as a collective journey through the zone of proximal development of the activity: It is the distance between the present everyday actions of the individuals and the historically new form of the societal activity that can be collectively generated as a solution to the double bind potentially embedded in the everyday actions. (Engeström, 1987, p. 174)

A typical sequence of learning activities in an expansive cycle is described by Engeström (1999b, p. 383) as follows:

Questioning – criticism or rejection of some aspects of accepted practice and existing knowledge.

Analyzing the situation – The analysis includes the symbolic and practical transformation of the state of affairs to discover the causes or illustrative mechanisms. The analysis relies on questions like "why" (explanatory). One type of analysis is "historical-genetic" and seeks to explain the situation in view of the origin and evolution. Another type of analysis is "real-empirical" and seeks to explain the situation to build a picture of internal relations.

Modelling – Formation of a newly explanatory relationship in a receptive and easily understandable model. This means an explicit, simplified model of a new idea that explains and offers a solution to the problematic situation.

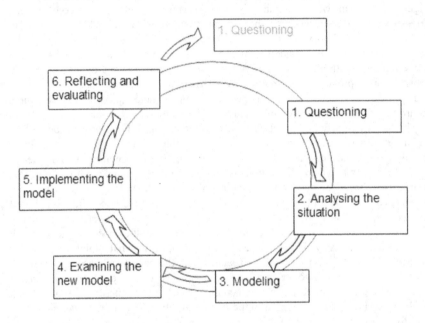

Figure 4. Sequence of epistemic actions in an expansive learning cycle (Engeström 1999b, p. 384).

Examining the new model – implementation, operation, and experimentation with this model so that its dynamic capabilities and limitations can be brought out.

Implementing the model – implementation of the model through practical applications, enrichment, and conceptual extensions.

Reflecting and evaluating – process evaluation and final acceptance in a new, stable form of practice.

The extension of activity theory and the development of expansive learning by Engeström offer a crystal clear view and a new methodology for organised learning in an effort to develop and organise new teaching practices. This perspective is most needed now, as the development of a new methodology of teaching should be able to fully exploit the potential and benefits of new technologies, not only in the classroom but in workplaces where there is no clear pattern. The pedagogical issues are important for promoting innovation with new practices that do not resemble the past (Rizzo, 2003).

In his five principles of activity theory, Engeström considers four basic questions that determine theoretical knowledge and the outcome of any course of learning.

1. What are the subjects of the learning process and how were they selected and defined?
2. Why do they learn? What makes them endeavor?
3. What do they learn? What is the content and outcome of learning?
4. How do they learn? What are the key actions or procedures?

Table 1. Matrix for the analysis of expansive learning basic questions (Engeström, 2001, p. 138).

	Activity system as a unit of analysis	Multi voicedness	Histori-city	Contradictions	Expansive cycles
Who is learning?					
Why do they learn?					
What do they learn?					
How do they learn?					

Engeström proposes the use of such an analysis for organized learning environments based on the motivation of individuals to develop various strategies which cause internal tensions and contradictions that form a momentum that will enable people to trigger a serious effort to achieve the objectives of learning (Rizzo, 2003). For the concept of expansive cycle, Engeström begins with individual concerns that question the accepted practice until now and gradually expands into a collective movement from the abstract to the concrete through specific procedures.

In the design of teaching materials for students in sixth grade of primary school, as described above, using Cultural-Historical Activity Theory, we incorporated data from the History and Philosophy of Science. The next section describes the key stages that scientists in the History and Philosophy of Science went through in order to achieve the integration of magnetic and electrical phenomena and the formulation of the theory of electromagnetism during the 19th century. These stages are the backbone of electromagnetism as taught in elementary school.

The History and Philosophy of Science and its Relationship with the Teaching of Science

It was the Greek philosopher Thales of Miletus (~624–547 BC), who circa 600 BC first made a written reference to the phenomenon of magnetism. From the times of

123

Thales and till the 16th C. electrostatic and magnetic phenomena were unified in the context of a "magic" idea and were supposed to be of the same nature. Their differences were pointed out during 16th C. by Gardano and Gilbert, and two fields of science were established: electrostatics and magnetism. From the 17th C. and up to 1830, scientists dealt with the question of whether electricity derived from different sources was of the same nature. During 1832–1833, Faraday successfully carried out a number of experiments to compare the ability of various electricity sources to produce the same effects (Seroglou & Koumaras, 2001).

The vitalist-animist model and the model of Gilbert are the first attempts to interpret magnetic and electrical phenomena, which are displayed macroscopically and characterized by (Guisasola et al., 2005):

1. The step from the unification of forces to their separation, that is, from the lack of distinction between the gravitational, electric, and magnetic forces to their differentiation into separate forces;
2. The step from a methodology of investigation which was not experimental or qualitative to one based on empirical evidence but not quantified (without a mathematical model);
3. A first attempt to collect data from experimental data.

Vitalist – Animist Model

I.   From the ancient Greek philosophers until 1600.

Ancient Greeks were historically the first to attempt a systematic collection of knowledge based on human reason, without limitations and prejudices. So Greek civilization left behind it the basis of a rational philosophy and so the origins of modern western science.

Some ancient Greek philosophers (Thales,[2] Plato,[3] Aristotle), in trying to explain the attraction of magnets and especially magnetite and electrum, gave different interpretations, but their interpretations had a common denominator, which attributed feeling to animate or inanimate matter. The model that had prevailed up to 1600 implied the existence of a hidden force which was embedded in iron and activated by the presence of a magnet or it was moved from magnet to iron. This hidden power is a feature of both amber and magnets and causes the tendency of iron and small objects to move towards the magnet or amber. In ancient and medieval times no distinction was made between gravity and electrical and magnetic forces (Guisasola et al., 2005). A second model by philosophers who believed that the magnetic force is due to some kind of "flow" that is emitted only from the magnet, or even by iron, was not accepted as widely as the Aristotelian view (Voutsina & Ravanis, 2007).

II.   Gilbert's model (1600)

In an era of rising humanism and the Renaissance (15th and 16th centuries) a very important new understanding of nature was born. Access to nature was now sought through experiment – a special, modern concept that was first shown clearly in Leonardo da Vinci (1452–1519). Experimenting meant setting questions to nature in relation to a prior theory formulated to examine whether the theory is confirmed or rejected by experiment. (Heisenberg, 1997, p. 190)

Gilbert was one of the first to implement this new attitude in the examination of magnetic and electrostatic phenomena.

Modern science and technology largely determines our present attitude towards nature, which is in contrast to that applied in previous centuries, during which natural philosophy expressed an anthropocentric view.

"Any knowledge which does not depend on external causes but only on itself is, for Aristotle and the Aristotelian tradition, knowledge through which the essence of man is realized" (Rossi, 2004, p. 49). This view, that knowledge is devoted to the search for truth, prevailed in ancient and medieval times and led to the belief that mechanical arts, manual work and, by extension, experimentation were lower forms of knowledge. In combination with the theocratic conception of nature, which is presented as a divine creation, any inquiry about the nature of the material world seemed absurd.

The rejection of these frameworks, although launched in the Middle Ages through individual efforts such as those of I. Filoponos, had no effect until the early 17th C., when the whole edifice of the Aristotelian tradition was brought into question. So by the early 18th century, the attitude toward human nature had changed radically. Delving into the details of physical processes, the researcher realized that it was possible to identify individual physical phenomena, a mathematical description, and a resulting "explanation" (Heisenberg, 1997, p. 13). This is the period of amateur science (1600–1800) (Butterfield, 1994). Unlike today, science was not pursued in universities, government, and industry. The participants were men who practised a profession, they were financially independent, and their main social role was associated with something beyond their interest in science (Woolgar, 2003, p. 40)

William Gilbert was Queen Elizabeth's doctor, but his contacts and discussions with sailors led him to deal with magnetic phenomena, as reflected in his treatise "De Magnete, Magneticisque Corporibus et de Magno Magnete Tellure; Physiologia Nova, Plurimis et Argumentis, et Experimentis Demonstrata (About the Magnet, the Magnetic Bodies and the Earth, the Mega Magnet – New Physiology Proven by Many Arguments and Experiments). It is very difficult to answer the question whether this is the last work of natural magic during the Renaissance or one of the first modern scientific works (Rossi 2004). Gilbert experimented but not in a systematic order: the only consistency present was his willingness to try anything that could be done with a magnet (Butterfield, 1994, p. 101).

Gilbert's model is characterized by a change in the methodology used in previous years. It is based on the results of empirical evidence, as opposed to the qualitative focus of previous centuries. Nevertheless, magnetism is understood as a characteristic feature of the objects, such as mass, volume, etc. Both earth and magnets have an intrinsic magnetism. Gilbert not only experimented with magnets, but suggested that the strength of magnetic attraction is the real cause of gravity and that it explains why the various parts of the world remain united (Butterfield, 1994; Rossi

2004). Gilbert's views on gravity took their place among the most prevalent ideas of the 17th century.

Nevertheless, his views were not accepted by scientists such as Kepler, Bacon, and Boyle. Eventually Galileo seemed willing to adopt the more general theories of Gilbert, but he accepted that he did not understand magnetism or its role in the workings of the universe and regretted that Gilbert was so limited as an experimenter and that he failed to express magnetic phenomena in mathematical terms (Butterfield, 1994, p. 143).

In contrast to Aristotle's experiments, which were based on experience, Galileo's experiments were based on an ideal world whose reality is but an imperfect realization. Galileo thought of experiments primarily as ideas with which to persuade others, and he was ready to announce his results with absolute certainty without being bothered to perform experiments (Westfall, 2006).

One of Gilbert's important innovations is the clear distinction he made between magnetic and electric action. He also introduced the term electric power, a term which will attain in the future a unique success (Binnie, 2001; Kipnis, 2005; Rossi, 2004). Electricity, a term that does not appear in his writings, is presented as an attraction that all small and lightweight items undergo (e.g., amber, jet, glass, resin, and sulphur) when rubbed (Rossi, 2004, p. 353). He made the versorium, which was a real electroscope.

Behind the elaborate and intelligent experiments of Gilbert lies a "magical-vitalist" perspective in which matter is not devoid of life or perceptual skills (Rossi, 2004, p. 353). The force of attraction exerted is always proportional to the quantity or mass of the body exerting the force: the greater the mass of magnetite, the greater the "pull" exerted by it on objects. This force is not exerted from a distance but by effluvia materiali – a visible action creates invisible fumes (Butterfield, 1994; Rossi, 2004).

III. The Fluid Model

The second main qualitative leap was from the vitalist to the corpuscular physical theory (from the 17th to the 18th century), characterized by Guisasola et al. (2005) and Kipnis (2005):

a) a step from the vitalist vision to the corpuscular. That is, magnetism is not considered anymore as an innate quality of some materials but as an invisible and unique fluid. Later on, the fluid model is abandoned and magnetism is defined by means of its effects and its operative definitions, finally making a clearly positivist interpretation;

b) a step from a qualitative to a quantitative methodology in which, despite its quantificational aspects, physical magnitudes are not expressed in mathematical terms.

a. The Single Fluid Model Of Aepinus (1724–1802)

As in Gilbert's case, this model was intermediate between the "single fluid" (the prevailing standard for the interpretation of electrostatic activity), which explained magnetic activity, and the model of "action from … distance," or Newtonian model (which started to become the scientific paradigm of the time) which undertook the interpretation of magnetic and electrostatic activity thereafter (Heilbron, 1979).

In the 18th century, the prevailing model was of one or two types of magnetic fluid. The main representative of the model for one type of magnetic fluid is that of Aepinus, who found that the magnetic poles are areas where there is a surplus or deficit of magnetic fluid. He offered an account of the constancy of the magnets and believed that the magnetic fluid is firmly connected to the pores of the material from which it emanates. He assumed that the magnetic fluid particles repel one another and attract particles of iron (Whittaker, 1910). Representatives of the model positing two types of magnetic fluids are Brugmans and Wilcke, who gave them the names north and south; Coulomb found that unlike electric, magnetic fluids cannot be separated, since the two magnetic poles cannot be isolated (Whittaker, 1910).

b. The Newtonian Model of Coulomb (1736–1806)

This model, though clearly inspired by Newton, interpreted the nature of magnetism by positing the existence of two fluids and thus claimed to explain why there are no monopoles (bodies with only one magnetic pole). The main contribution of Coulomb, in the study of magnetism, was the development of a mathematical model which purported to explain the law of magnetic force. He introduced a quantitative method, similar to that used by Newtonian mechanics, and defined its functionality through dynamic effects and the concepts of the magnetic pole and magnetic force (Heilbron, 1979). In this way, the theory of Coulomb applied Newtonian mechanical formulas to magnetism.

IV. The "Big Leap Forward:" The Appearance of Electromagnetism in the 19th Century (Øersted).

A new area opened in 1820, with an impressive statement by Hans Christian Øersted that electricity diverts a magnetic needle, which was originally oriented parallel to an electric cable. Already in the 18th century, it was suspected that there is a link between electricity and magnetism and several experiments were made towards this direction. Now, in the early 19th century, Øersted's experiment verified this link, as testified to by the magnetic effects of electricity, and paved the way for the emergence of electromagnetism. The culmination of his experiments was the breakthrough of 1820: just as a magnet, electricity had the same influence on a magnetic needle, and thus electromagnetism was born.

Øersted's experiments showed the existence of the "power cycle," i.e., of a torque that causes the magnetic needle to rotate. This movement was unexplained by the Newtonian view. Øersted's article was mainly qualitative but created an opening for the study of electromagnetism (Segrè, 2001). Soon after, in the autumn of 1821, came the discovery by Ampère of attraction or repulsion between two pipes through which electricity is passing. The discoveries of Øersted and Ampère contributed to the correlation between magnetic phenomena and electricity (Seroglou et al., 1998).

127

In attempting to bring electromagnetic forces into the province of Newtonian mechanics, Ampère and others constructed a mathematical representation of them as actions at a distance. Continuous-action phenomena, such as fluid flow, heat, and elasticity, had all recently been given dynamic analyses consistent with Newtonian mechanics. The initial presumption behind these analyses was that, at the micro-level, there are forces underlying action-at-a-distance in a medium and it is these forces which are responsible for the macro-level phenomena (Nersessian, 2002).

The discoveries of Ampère and other researchers in the early days of electricity were expressed in pure mathematical language, so they were representing the ultimate development of Newtonian physics based on the laws of forces between two current elements, or two point loads on the move (Segrè, 2001).

The main model for the study of magnetism in France and Germany in the 19th century was Newton's model of action at a distance. Main representatives are Ampère and Weber. Ampère believed that magnets are created by tiny circular molecular currents which come about through the interaction between currents and magnets; magnetism is simply an interaction between currents (Guisasola et al., 2005).

Scientists of continental Europe followed the trail of Coulomb and Ampère in expressing electromagnetic forces in a differential typology. "They presented the results of the attraction of repulsion and induction as remote interaction of different particles motivated by the Newtonian law of gravity and the expansion of the inverse square relationship to the electromagnetic forces" (Gillispie, 1994).

V.   The Field Model

After the discovery of Øersted about electromagnetism, and the arrival of field theory of Faraday, two ontological views coexisted in the scientific community in the 19th century: "action at a distance" and "field theory." They didn't run in parallel, but were often isolated from one another and they formed the basis for all subsequent theories that have arisen.

a.   The Field Model of Michael Faraday (1791–1867)

Scientists in England followed a different path. Faraday was the first to demonstrate the reciprocal relationship between electricity and magnetism. After following a series of experiments that created electricity by moving magnets, he offered the idea of a dynamic line, which he actually envisaged as being real in the physical condition of space and defined the concept of power as a key entity in space. In this way, he is the precursor of field theory (Gillispie, 1994).

Faraday had made the hypothesis that the lines of force which are formed when iron filings are sprinkled around magnets and charged matter indicate that some real physical process is going on in the space surrounding these objects and that this process is part of the transmission of the actions (Faraday 1835–55). In 1830, Faraday successfully carried out several experiments to settle the matter and concluded this long scientific argument. (Seroglou et al., 1998). Faraday was convinced that the relationship between electricity and magnetism had to be extended, and if the current created a magnetic field, then the magnetic field should be able to create electricity. He had grasped the essential point that, to produce current, a pipeline had to interrupt the lines of magnetic forces (Segrè, 2001).

b. The Field Model of W. Thomson (1824–1907) and J. C. Maxwell (1831–1879).

This model was introduced for the first time by Thomson and then by J. C. Maxwell, who relied on the existence of ether and engineering to express mathematically the concepts of Faraday and make them more understandable. William Thomson introduced a method of investigation known as "the method of analogies," understood as the following: "the mathematical analogy between fluid of liquid, heat and electricity and magnetism implied a mathematical similarity not a physical one …"; the similarity occurs between the mathematical explanation of this phenomena, not between the phenomena themselves (Nersessian, 1995, 2009).

Table 2. Scientist-scientific model and activity

Greek ancient philosophers	Vitalist–Animist Model	1. From toys with magnets in magnetic phenomena
William Gilbert (1544–1603)		2. From the attraction of magnets to the attraction of other bodies
Franz Aepinus (1724–1802) Charles-Augustin de Coulomb (1736–1806) Charles du Fay (1698–1739) Benjamin Franklin (1706–1790) Alessandro Volta (1745–1827) Luigi Galvani (1737–1798)	The Fluid Model	3. From animals' electricity to batteries' construction
Hans Christian Öersted (1777–1851) André-Marie Ampère (1775–1836)	The "Big Leap Forward:" The appearance of electromagnetism in the 19th century (Øersted).	4. The new discoveries that changed our world: From electricity to magnetism: Øersted's experiment
		5. The new discoveries that changed our world: electromagnetism.
Michael Faraday (1791–1867) W. Thomson (1824–1907) J. C. Maxwell (1831–1879)	The Field Model	6. The new discoveries that changed our world: Faraday's experiments that led to electric motors.
		7. The new discoveries that changed our world: Faraday's experiments that led to electric generators.

Maxwell, on the other hand, connects Faraday's ideas to the mathematical analogies of Thomson. Basically, Maxwell's objective was to discover an appropriate model which could explain the basic mechanism of electromagnetic phenomena. He chose Thomson's vortex model. This model allowed him to work out a group of laws which explained all the magnetic phenomena (at a macroscopic level) in a formal way, as well as optical physics. Later he abandoned the mechanical model because it created too many difficulties, and he was left with his equations (what scientific historians call the "The Operative Interpretation"). Hertz's discovery of the waves which are called after his name was the experimental confirmation of Faraday and Maxwell's field theory (Guisasola et al., 2005). Maxwell expressed in mathematics the field concept of electromagnetic forces, formulated the laws of a non-Newtonian dynamic system for the first time, and constructed this novel representation by abstracting and integrating constraints from what was known of continuum mechanics and machine domains and from the new domain of electromagnetism into a series of models and formulated the quantitative relationships among the entities and processes in these models (Nersessian, 2005).

Using the stages of the development of scientific thought on electromagnetism, we constructed corresponding activities to be taught in primary education as shown in Table 2. These activities follow the structure of an expansive cycle and are presented through a computer. Each step of the expansive cycle is a "screen" in which students can work to complete the activity. To be presented effectively and to draw the interest of the students to the data from the History and Philosophy of Science, we use different teaching strategies as described in the literature; they are presented in Chapter 7.

Teaching Strategies Using Data from the History and Philosophy of Science

Various proposals concerning the contribution of the history of science to science teaching were drafted in the early 21st century (Seroglou and Koumaras, 2001). However, integration of elements of the history of science into science teaching seems to acquire a different meaning which is related to the kind of transformation scientific knowledge undergoes when it becomes a school subject (school knowledge).

According to Monk and Osborne (1997), the role of epistemology is crucial to the incorporation of History of Science because the answer to the question of "how we know" is an important aspect of our account of science and provides evidence for our ontological commitments. Furthermore, in the context of science education, scientific epistemology is a central concept – that is, "to tell how to distinguish between justified and unjustified beliefs" is an essential critical skill required for participation in any scientific discourse. For Monk and Osborne, the introduction of the History of Science "will continue to remain more talked about than taught as long as the materials that teachers are provided with have an additional character focusing on the context of discovery, rather than the dominant perspective of mainstream science teaching the concerns of which are in accordance with the products of epistemological justification and the methodology of science (Monk & Osborne, 1997).

Matthews (1994, 1998), along with Stinner and Williams (1998), support the view that the history and philosophy of science should be in the curriculum and proposes that it should:

- be within the context of students' lives, create incentives, inform, and be relevant to the interests of students.
- incorporate the view that the historical context helps students understand that scientific knowledge is not permanent.
- give a human dimension to science and link students with the personal, moral, cultural, and political interests of the people.
- create direct experience for the students who grow up in an environment where information is provided electronically by new technologies, which bring great benefits but also great risks.

To the above we could also add the help which is provided to the students, so that they themselves can: (a) evaluate the landmarks in the evolution of ideas and technologies, and (b) discern what is the important information within the plethora of information available due to the growth of electronic information and especially the Internet.

The main benefit of learning the History of Science consists in the active participation of students. Seroglou and Koumaras (2001) argue that teaching material, which has been designed to facilitate the study of the work of the scientists who helped change scientific ideas and led to the scientific theory acceptable today, could include activities designed and inspired by:

1. Historical experiments with highly visible features that helped change scientific ideas in the history of physics.
2. Abstract reasoning which helped change scientific ideas in the history of physics.
3. A combination of experiments and thoughts that helped change scientific ideas in history.

The transition of thought from the world of senses to a higher level, where the main role is played by concepts, is essential to carrying out activities in the teaching process within the classroom.

Within such a context, data from the History of Science can be introduced in the following forms (Stinner et al., 2003):

1. Story Line We create a story line (perhaps historical or conceptual) that will be activated and will emerge through this central idea/concept. Identifying an important event linked to a person or persons and finding opposing couples or conflicting characters or events that may be suitable for inclusion in history (Stinner et al., 2003). In the development of a scientific concept, scientists sometimes reject the ideas of earlier scientists and develop new ones. Sometimes scientists interpret events differently and extend or modify previous theories. This scientific process continues as concepts develop throughout the course of history.

The stages in the development of scientific knowledge can be constructed – throughout the course of history – as a story line. Stinner and Williams (1998) confirm that the substance of discussions among scientists could help teachers to participate in similar discussions with their students.

Stinner and colleagues (Stinner & Williams, 1993; Stinner et al., 2003) suggest using the story line approach in order to attract student attention and engage their imagination. In the development of a scientific concept, scientists sometimes reject the ideas of previous scientists and develop new ones. Sometimes scientists interpret phenomena differently and extend or modify previous theories. This scientific process goes on as concepts develop throughout history. Stages in the development of scientific knowledge throughout history can be constructed as a story line. Stinner (1995) emphasizes the importance of the grade level of students in the use of a story line for science education. In the elementary school, science stories should be based on student imagination, in middle school science stories they should be based on history, and in high school years popular science literature related to the content should be used to create a teaching context. Stories of science should be developed for students in the early years and middle years.[4]

These story lines may also include the major historical events which occurred during the life of those scientists and influential thinkers of the era, in different fields such as philosophy, social sciences, politics, and literature. It is sufficient if we spent only a few minutes on any different scientific approach followed by other scientists. This creates a connection with the scientists we are interested in, who are the best of their era so that the students may create the right connections. In the context of a story line, any and all following strategies can be used for teaching science.

2. Case study. A case study is a smaller part of a large context of a central idea/ concept and is designed by a team of three students called to present the case study to a group, or rather to the audience of the class (Allchin, 1997; Irwin, 2000; Stinner et al., 2003; Bevilaqua and Giannetto, 1998). Each student undertakes to present one of the following sections:

1. *Historical context:* Present scientific ideas of the particular historical period and show their relation to the matter under consideration.

2. *The experiment and the main ideas/concepts:* Present the basic idea or concept and the empirical data arising from history that are essential in the case under consideration.

3. *Impact*: Present the effects of the central idea or empirical evidence in scientific literacy and the teaching of science. Students respond to the following questions: Where do the concepts fit into the science curriculum? How would one present these concepts/ideas/experiments in the classroom? What are the diverse connections of the concepts under discussion? (Stinner et al., 2003, p. 620)

3. Reconstruction materials (replicas) and reproduction of historical experiments. The idea that an experiment can change the course of science is an interesting claim and it is a challenge for the students to reproduce it. It would be ideal to use equipment similar to that used originally by the scientist. Using the same construction procedures and operation of devices which originally the scientists used in their research accomplishes the following, according to Heering, (2003):

– Overcomes the apparent linearity of the development of scientific knowledge.
– Puts students in an unusual situation: The experimental situation is "open," meaning that they do not know what may result from their experiment. The experience is different in the experiments used in the teaching of science. Neither is discussed in textbooks nor are the results of experiments essential according to modern theories.

And finally, the explanation for the rejection of unsuccessful experiments is a much more complicated one. Therefore these examples can allow students to overcome the epistemological position of a single experience. The lack of discussion of unsuccessful experiments can produce a perception of the nature of science that is contrary to original intentions (Heering, 2005).

4. Dramatisation and role play. Theatrical plays based on the controversies involving great scientists (e.g. Copenhagen, involving Heiseberg and Niles Bohr) can be staged by students and presented to audiences. For example, the plays of: Brecht, "The life of Galileo"; Golding: The Physicists; Kipphard: In the Matter of J. Oppenheimer. Unfortunately, these works have not been translated into Greek, except for Brecht's "The Life of Galileo," and it is therefore difficult to transfer some plays into Greek. Besides, such a theater experience is applicable to a more mature audience and, therefore, it is suitable mainly for university classes.

Role play activities in educational settings serve multiple purposes. Students engaged in such activities may take the role – in this case a relevant selection of physical properties mapped onto physiological or social elements – of physical entities like atomic particles or fields. This method is more common in younger children's science education in primary school. Its characteristics may be used to explore the role of models in science. Within an HPS approach, it may serve to present the changing nature of modelling phenomena through history, for example, the differences between fluid-based and particle-based modelling of electricity (Henke et al., 2009). Science drama is expected to provide a humanistic aspect to authorized scientific knowledge by eliciting emotional and active participation. In addition, science stories must consciously incorporate a "scientific element" and a "humanistic element." Even for the simple retelling of "Eureka stories," the crafting of the story is a humanistic, creative process. We can invent stories, but they must be well placed in history (Stinner, 2007).

Yoon (2006) suggests three ways of using science drama in the classroom. Firstly, the setting of the scene should be designed and in the process it provides a useful context for learning. Secondly, drama making is a learning process on its own. The intention is for a great deal of learning to occur during the preparation of science drama. Thirdly, drama as a representation of aspects of natural philosophy offers an assessment opportunity.

5. Biographies of scientists. Stories about the personal lives of scientists stimulate students' interest in science. Discussions just of the methodology followed by each scientist usually lessens students' interest (Seker & Welsh, 2006). For example, Aristotle and Galileo used different scientific methods to produce scientific knowledge. Aristotle based his conclusions on empirical observation, while Galileo performed experiments. A teacher can show how short biographical passages relate to the content being taught.

Undoubtedly there are biographical sources accessible to teachers, but caution is necessary when they are used in teaching practice. It is possible, for example, for children to hear the story of Archimedes' "Eureka" and form the view that scientists spend much time in meditation, trying to solve problems by working on their own (Stinner et al., 2003). We need, therefore, stories from the personal lives of scientists to be directly associated with the content being taught.

6. *Thematic narratives.* This approach identifies general issues that transcend the boundaries of individual disciplines and may have interdisciplinary and human interactions. These issues transcend individual disciplines and often link important aspects of human activity (Dagenais, 2003; Stinner et al., 2003).

7. *WebQuest, simulation of historical experiments and site design.* Masson and Vázquez-Abad (2006) propose a way of integrating data from the History and Philosophy of Science in teaching science to achieve conceptual change by introducing the notion of a historical micro-world, which is a computer-based interactive learning environment. Also, in our previous work we presented educational software (CD-rom) about the life and work (for science) of Archimedes (Kokkotas et al., 2003; Stamoulis et al., 2006), which can be used for teaching simple machines in elementary education.

WebQuest is a lesson script, a learning activity that focuses on enabling students, and it is oriented to research. It is a problem-solving activity where students can use various information sources (Web, educational software, contract forms, etc.) (Dodge, 2001; WebQuest Resources). The term WebQuest was first introduced in 1995 by Dodge to describe structured exploratory activities of pupils or students, where most of the information was drawn from the worldwide web. The means used can be hypertext information, databases, electronic data or interviews, discussions, etc., printed conventional material from books, magazines or newspapers, and face-to-face interviews with experts, students can choose from the sources available to them. Biographical data, simulations, anecdotal descriptions of historical experiments, and scientists' authentic texts can be utilized to create a WebQuest to learn the content of science using data from the History and Philosophy of Science.

Conversely, students can collect material to clarify the content, use elements of the story related to the negotiated notion, and upload to a website. This act requires a higher function of thought (organization, classification, abstraction, judgment) and is therefore quite a productive activity for students. However, it needs to be supervised by a teacher.

8. *Confrontations.* We tend to think that modern science can resolve most issues, but the truth is that the science of the 20th century is full of conflicts; some have

been resolved completely and some only partially, while others remain unresolved. Sometimes there are competing theories that seek to lay the foundations of new issues, such as the science of the 18th century, electricity, and the new chemistry of Lavoisier and alchemists, but more often scientific conflicts involve the settling of rival theories (Bevilaqua & Giannetto, 1998; Stinner et al., 2003).

Moreover, the study and use of scientists' conflicts (Galvani and Volta) enriches the teaching of science, motivates students, and makes learning more meaningful. Students understand scientific ideas, scientific methodology, and the nature of science much better through the lives and work of scientists (Malamitsa, Kokkotas, & Stamoulis, 2005).

9. Vignettes. The smallest unit of presentation of historical material may be a short story, carefully selected and linking the concepts and ideas to the study and the interests of students (Stinner et al., 2003). Vignettes of the History and Philosophy of Science are specific activities that can be examined in the context of the activity theory (Engeström, 1999a). In this case, evidence from the History of Science provides tools of mediation, through which social interaction and cooperation can be enhanced to achieve the learning objectives. Carefully selected small vignettes are incorporated organically in worksheets and contribute effectively to mediation in achieving the learning objectives of students. The worksheets can be integrated in a broader context of the story line for a teaching concept.

10. Dialogues. Dialogues between representatives of opposing theories can be presented in class, for example, as dialogues between Copernicus and Aristotle, Priestley and Lavoisier, and others (Stinner et al., 2003). As students try to find examples to support their theory, they begin to understand the process of science. The students voluntarily participate in the discussion. They prepare for discussion by collecting information about the scientist they have chosen to present in the class. The discussion in the classroom is quite often lively and often brings out quite intelligent examples (Dagenais, 2003).

11. Historical thought-experiments. Thought experiments devised by scientists during their research can be tested or verified with our resources. Hypothetical experiments have played an important role in the history of science. Evidence of this is their use by Galileo, Leibniz, Newton, and Carnot and, in the 20th century, by Einstein, Schrödinger, and Heisenberg. According to Matthews (1994) hypothetical experiments can be distinguished into those that are destructive of already accepted theories or conceptual schemes and those that are constructive, or that support new or old theories. Presumptive experiments enable teachers to assess the extent to which students understand the fundamental principles of an axiom; they employ the mind and reveal what the student thinks about the concepts being investigated (Matthews, 1994).

12. A variety of teaching tools such as design and layout of posters, discussions on the occasion of a historical person, etc. This approach shows a great deal of promise for facilitating the use of history of science in science teaching as well as the strong influence that the study of the history of science may have on researchers coming from a variety of fields (science education, history of science, philosophy of science, information and communication technologies in education, art studies, social studies, etc.).

TOWARD TECHNOLOGICAL SUPPORT FOR DESIGN ACTIVITIES

Activity theory puts the general trend for the design of interaction outside of computers and focuses on understanding technology as part of a broader goal of human activity (Kaptelinin & Nardi, 2006). Present times are characterized by the ability to communicate with people anywhere in the world. The introduction of ICT (Information Computer Teaching) in education is now a necessity and a high priority for the educational systems of all countries. However, successfully integrating digital tools into everyday practice is a complex endeavour. The use of technology for educational purposes is currently a common practice. This practice is constructed by society –a product of the values, knowledge, and skills of people. Inevitably, therefore, it is influenced by the social and cultural context in which it develops. Closely connected with this view is the idea that the impact of technology in facilitating students' access to learning is also determined by the pedagogical knowledge and skills of teachers. Technology enables teachers to extend the validity of theories (pedagogical and psychological for learning) the absence of which leaves teachers in mediocrity (Elmore, 2004).

Within the field of social studies computers have served dual roles, as important instructional tools and as objects that have affected the political, social and economic functioning of society. However, the extent to which this potential is being fully realized in the classroom has not been sufficiently explored. Computer-based learning has the potential to facilitate the development of students' decision-making and problem-solving skills, data-processing skills, and communication capabilities. By using computers, students can gain access to extended knowledge links and broaden their exposure to diverse people and perspectives. "According to Vygotsky, whoever adopts the view that art is a means of pleasure and enjoyment, it is likely to encounter strong competitors in sight of the first delicacy to find the child in the way" (Dafermos, 2002). This view of Vygotsky is directly applicable today to computers and their use by students. The impressive design of modern environments, easy access to a plethora of exciting things (games, music, movies, etc.) and information can easily be "treats" for young students and lure them. What we need now is not for the computer to be a delight and pleasure to the child but to provide the necessary impetus to transform the "lower" forms of mental energy into "higher" forms.

To achieve learning, students must be engaged in their learning goal; the computer becomes the intellectual tool that will attempt to mediate the conquest of knowledge. In other words, when students work with computers, they enhance the computer's capabilities and then the computer enhances their thinking and learning. Like carpenters who use tools to build things without being controlled by their tools, so students should use computers as tools to achieve their own goal and to support learning and not be controlled by them and directed in other directions (Jonassen, 2000).

Humans not only created tools and technical resources to subjugate the forces of nature but they also created psychological tools that help to regulate and control mental activity (Dafermos, 2002). It is therefore necessary to escape from a techno-centered prospect where the computer is the focus for understanding technology as part of human activity (Kaptelinin & Nardi, 2006) in the direction that the computer is a tool for mediating the interaction between man and environment. Kozma (2003), in his introduction of technology and innovation in education, defines five criteria for innovation:

1. Changing roles of teachers and students, regarding the objectives of the curriculum, assessment practices, educational material, and infrastructure.
2. The essential role and added value of teaching practices.
2. The relationship of innovation to the achievements of students.
3. The capacity of sustainability and the possibility of spreading innovative practices from one classroom to another in local, national, or international levels.
4. The integration of national issues such as local and cultural particularities.

The first attempts to introduce ICT in education in the 1970s relate mainly to development of teaching systems using the computer (Computer Assisted Instruction CAI or Computer Assisted Learning CAL) and to training programs and "drill and practice." These programs were based mainly on the behavioural theory of learning. This methodology, however, cannot promote complex thinking, problem solving, and the transfer of skills acquired by students in other similar environments (Jonassen, 2000).

The most sophisticated form of CAI is an intelligent tutoring system (ITS), sometimes referred to as intelligent CAI. ITSs were developed throughout the 1980s and 1990s by artificial intelligence (AI) researchers to teach problem solving and procedural knowledge in a variety of domains. What ITSs add to tutorials is intelligence in the form of student models, expert models, and tutorial models. Expert models describe the thoughts or strategies that an expert would use to solve a problem. How the student performs while trying to solve the problem in the ITS (captured in a student model) is compared with the expert model. When discrepancies occur, the student model is thought to have bugs in it, and the tutorial model diagnoses the problem and provides appropriate remedial instruction. ITSs have more intelligence than traditional tutorials and so can respond more sensitively to learners' misinterpretations. [...] In the 1980s, microcomputers proliferated, so educators (as they had with most other technologies like radio, film, and television) began grappling with how to use them. The unfortunate result of their deliberations was that most educators felt that it was important for learners to learn about computers. So, we taught students about the hardware – components of computers. And because useful applications were not available, we taught students how to program the computers, too often using BASIC. [...] It is a mistake to believe that if students memorize the parts and functions of computers and software, then they will understand and be able to use them. (Jonassen, 2000, pp. 6–7)

According to Kaptelinin and Nardi (2006), whose contributions on activity theory have been influential in the domain of Human Computer Interaction, activity theory is a conceptual framework that enables people "to bridge the gap between motivation and action [and] provides a coherent account for processes at various levels of acting in the world" (Kaptelinin & Nardi, 2006, p. 62). The dialectical relationship between semiotic and technological spaces leads us to consider the concept of functional organs, which is viewed by Kaptelinin and Nardi (2006) as "a key concept of activity theory from the point of view of interaction design" (Kaptelinin & Nardi, 2006, p. 64). According to these authors, "functional organs combine natural human capabilities with artefacts to allow the individual to attain goals that could not be attained otherwise." To create and use functional organs, individuals need a range of competencies.

Tool-related competencies include knowledge of the functionality of a tool, as well as the skills necessary to operate it. Task-related competencies include knowledge about the higher-level goals attainable with the use of a tool and the skills of translating into the tool's functionality (Kaptelinin & Nardi, 2006, pp. 64–65). In addition to creating and using functional organs efficiently, individual subjects also need what Kaptelinin and Nardi call metafunctional competencies, which "enable [the subject's] understanding of how to use functional organs (such as knowing tricks and work-around), recognise their limitations, and know how to maintain and troubleshoot them" (Kaptelinin & Nardi, 2006, p. 218). The above competencies are also required of lecturers who want to efficiently integrate the institution ICT into their teaching practice.

Taking an ICT-mediated lesson in a school as an activity system, the subject is the student and the object is to understand the relationships among the variables found in an ICT-mediated simulation package. A pool of ICT and non-ICT tools, including the simulation package, mediates the interactions between the subject and object. The student belongs to a community consisting of his/her classmates, teachers, and ICT staff mediated by rules and division of labor. The rules include the school disciplinary rules and more specific ones like the procedures necessary to run the simulation program. For the division of labor, the student plays the role of the scientist, by gathering, representing, interpreting, and analyzing data, whereas the teacher takes on mediator role where he/she questions, clarifies, and summarizes to support students' understanding of the relationships among the variables under study (Lim, 2007).

Activity theory seems to be a natural fit in the HCI domain due to its being based on the central concept of the mediating role of tools in human work. However, activity theory, with its long history, wide-ranging complex, and rich approach and theoretical focus, has not lent itself to easy application in the HCI field. Activity theory provides no step-by-step methodology (Duignan et al., 2006).

Applying Activity Theory in Designing an ICT-Based Instructional Package. In this study we propose Engeström's expansive learning as a basis for planning and analyzing activities in science teaching. An activity may be the teaching of a concept of science. For example, let us introduce the concept of electromagnetism in elementary school. This activity is divided into activities that have to do with the creation of the concept of electromagnetism. So for each element which is a prerequisite for conceptual understanding by students of the concept of electromagnetism, there is an activity planned as a teaching/lesson.

For the design of each activity we follow the concept of the expansive cycle, which starts with the accepted practice of an action and leads progressively through the resolution of conflicts arising between the elements of activity, but also among the participants in the activity, to implement new practices. The integration of the cognitive structures of students' concept of electromagnetism is the goal of the instructional intervention that is the object of activity. The results of the activity are perceived through the potential of students to use the concept of electromagnetism or elements of the concept to solve problems of everyday life. The tool that mediates the subjects/students in the performance of the activity, the computer, is chosen as a natural tool, along with data from the History and Philosophy of Science as an intellectual tool.

The rules are set by the operation of the class and confine students to the tasks. The activities in this phase of work are addressed and take into consideration the community of the classroom. They are not extended to the family and the broader sociocultural environment of students for labor-saving. The division of labor is linked to the work of students both as individuals and as a group of a few students or as a group of the whole class. Then we describe the application of cultural-historical activity theory in the teaching intervention and the dialectical relationship between elements of the history of science, technology, and learning through an adapted version of Engeström's expansive activity model. Figure 5 shows the activity of electromagnetism as a drawing with the help of third-generation activity theory. The basic concepts of electromagnetism are taught in separate activities, which are part of the system activities with the common goal of understanding the concept of electromagnetism.

First activity – From toys with magnets in electric and magnetic phenomena. The teaching of the concept of electromagnetism begins by introducing students to the magnetic phenomena and their important function in the magnetism of the earth and the construction of the compass. The outcome of the activity is the use of the compass by students for their orientation and the application of the methodology of the experimentation that was first introduced by Gilbert. The students/subjects of the activity first classify materials with the help of the computer, and then with real objects of various materials both magnetic and nonmagnetic and then learn about the magnetism of the earth and its importance in the construction of the compass.

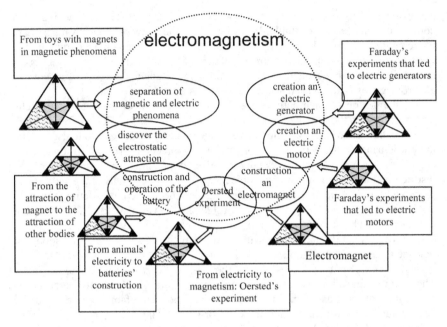

Figure 5. Applying the activity theory in the case of electromagnetism.

This activity is addressed in the class with the intention of achieving the target, a target mediated by a corresponding software and data from the history of science. Students are aware of the attraction of magnets and the model of the earth as a magnet and its importance in the construction of the compass and the orientation of people on earth. The rules and restrictions imposed on the operation of the class and, of course, the participation of each student or group of students in conducting activities, complement the other elements of the activity.

The structure of the activity follows the design of the expansive cycle of Engeström, whereby the first stage consists of a simple power function and creates an incentive and an analysis of the current situation while the scientists have only an empirical study of magnetic phenomena.

The analysis of the situation continues in the second stage, in which students are aware of the efforts of the scientist Gilbert to study magnetic phenomena scientifically. Also at this stage, the role of the earth as a magnet is introduced.

In the third stage of modelling tool which is mediated by the computer, students construct the model of the earth as a huge magnet and experiment.

The evaluation model is effected in the next stage of activity in which students name the poles of the magnet and utilize it in the construction of the compass. Acceptance of the model and the methodology of Gilbert, who studied magnetic phenomena scientifically, seems to help students evaluate the importance of science in developing scientific thinking and experimentation.

This activity is the basic unit of planning and analysis. The higher level of collective activity is directed before the goal of understanding the operation of the compass and magnetism of the earth.

The historicity of the activity allows students to conceive the initial attraction of the magnet to specific materials and the behavior of the earth as a magnet and then to externalise their knowledge by building themselves a compass, by evaluating scientific methodology, and by avoiding myths and beliefs in the interpretation of natural phenomena.

Table 3. The design of the first activity according to the expansive cycle

Questioning	Introduction of the topic (magnetism in ancient Greece)
Analyzing the situation	Students classify various materials into two different categories: those that are attracted to a magnet and those that are not.
Modeling	The model of earth as a magnet.
Examining the new model	Students experiment with the model of the earth as a magnet and know intuitively the dynamic lines.
Implementing the model	Students know the basic application of magnetism of the earth by creating and experimenting with a compass and naming the poles of a magnet.
Reflecting and evaluating	Students discuss the importance of the compass in the development of traveling and discovering the new world.

Second activity – From the attraction of the magnet to the attraction of other objects. In the second activity, students discover electrostatic attraction between opposite charged objects and the repulsion between objects charged with the same charge and recognize the symbolism of the load in positive and negative forms.

This activity is addressed in the classroom; achievement of the goal is mediated by the construction of Gilbert's versorium and the efforts of other scholars such as Du Fay and Franklin, who completed the study of the nature of electricity.

The activity begins with the question of whether there are other bodies in which there is an attraction. In the first stage of the activity there is an incentive to analyze the situation. The students know from everyday life that when the plastic cap of their pen is rubbed, it attracts small pieces of paper. Then they test the model of Gilbert's versorium on their computer. They test plastic material by rubbing it with a woolen cloth and a glass object by rubbing it with plastic, thus distinguishing the two types of electricity.

In the next stage of integration of the model, students are asked to use their own materials to construct the versorium and experiment. While in the process of accepting and evaluating, the students describe and interpret the function of the versorium using the model on the computer.

Table 4. The design of the second activity according to the expansive cycle.

Questioning	Incentive
Analyzing the situation	Students know that apart from magnet's attraction, there's also the electrostatic attraction among bodies
Modeling	Students run the versorium model on their computer.
Examining the new model	Students construct a versorium with real materials and experiment.
Implementing the model	Students know Du Fay's results for the two types of electricity and their denomination.
Reflecting and evaluating	Students give their own interpretation on the operation of the versorium and the nature of electrical loads.

Third activity – From animal electricity to construction of battery. One of the elements necessary for developing the concept of electromagnetism is the construction and operation of the battery, which can be approached in its historical dimension, namely with the construction of the electric battery by Volta. So we designed such an activity, whose outcome is battery usage by students in their everyday life. In this activity the students/subjects of the activity construct a simple battery of Volta with simple materials – the object of the activity mediated by cultural tools of the computer and the famous confrontation between Volta and Galvani on the nature of electricity, i.e., if that electricity is of animal origin or due to an electron flow due to potential difference. This activity is addressed to the class and the target's achievement is mediated by the corresponding software that we developed, through which students become aware of the controversy scientists engaged in over the creation of the model of the battery as well as the construction of a simple battery. The rules and restrictions imposed by the operation of the class and, of course, the participation of each student or group of students in conducting activities, complement the remaining elements of the activity.

The structure of the activity is designed to follow the expansive cycle of Engeström, whereby the first stage is a simple power function and creates an incentive to analyze the current situation, in which scientists can collect electrical loads and use them whenever needed. The analysis continues in the second stage, in which Galvani gives his own interpretation of electricity and the emerging confrontation with Volta. In the third stage of the modeling the tool of mediation is the computer, students construct a virtual model and bring it into operation. The battery of Volta is a fact and, in the next stage, students are asked to accept and integrate the model of Volta's battery, creating a potential difference between two metal electrodes and constructing a battery of simple ingredients (lemon and electrodes of iron and zinc).

The evaluation and acceptance of the model is amplified in the next stage of activity in which students compare a battery that they use in their everyday life with a battery constructed by Volta, which they constructed first on the computer and then with simple materials. This activity is a basic unit of planning and analysis. The higher level of collective activity is directed before the goal of understanding the operation of the battery.

Table 5. The design of the third activity according to the expansive cycle.

Questioning	Incentive
Analyzing the situation	Students know the views of the two scientists (Galvani and Volta) as well as their antagonism.
Modeling	Students construct a model of a virtual battery on the computer screen.
Examining the new model	The students bring into operation the model they constructed.
Implementing the model	Students create a potential difference between two metal electrodes by constructing a battery with simple materials (lemon and electrodes of iron and zinc).
Reflecting and evaluating	Students compare a battery used in their everyday life with a battery constructed by Volta, as they also constructed a virtual one first on the computer and then with simple materials.

This structure and comparison of the batteries they use every day is incorporated into their cognitive structures and allows them to internalize the operation of the battery and use it easily in their everyday life. The activity about the opposing scientists, Volta and Galvani, also reveals the contradiction that allowed both scientists and students to create electricity

Fourth activity – From electricity to magnetism: Øersted's experiment. The concept of electromagnetism is accomplished by teaching Øersted's experiment, which underlies the relationship between electricity and magnetism. The outcome of this activity is for students to understand the relationship between electric and magnetic phenomena. The students/subjects of the activity test the experiment's model on the computer and then construct it themselves with actual materials.

This activity is addressed in the class, and the objective is mediated by corresponding software and the historical experiment of Øersted and other elements from the history of science. The rules and restrictions imposed by the operation of the class and, of course, the participation of each student or group of students in conducting activities complement the rest of the elements of the activity.

The structure of the activity is designed to follow the expansive cycle of Engeström, whereby the first stage is a simple power function and creates an incentive for analyzing the current situation, which is that scientists have been trying for many years to discover if there is a relationship between electricity and magnetism in order to achieve integration of these two concepts.

Table 6. The design of the fourth activity according to the expansive cycle.

Questioning	Incentive and analysis.
Analyzing the situation	Students know the efforts of scientists and discover the relationship between electric and magnetic phenomena.
Modeling	Students virtually construct the model of Øersted's experiment on the computer screen and test it.
Examining the new model	The students bring into operation the model on the computer, along with the one they constructed with real materials.
Implementing the model	Students apply what they learned in the construction of the experiment on a problem of everyday life in interpreting the disorientation of a compass from lightning.
Reflecting and evaluating	Students accept the relationship between electricity and magnetism and try to convince the captains of the ships that the electricity in the lightning is what caused the disorientation of the compasses.

In the second stage of the activity the students learn from the original description of the experiment by the scientist himself; they are puzzled and then acknowledge that great achievements in science occur over a period of time, after having been preceded by several failed attempts to reach the goal.

In the third stage of the modeling tool mediation computer, students construct the model of Øersted's experiment on the computer and monitor its progress.

The model's evaluation is effected in the next stage of activity, in which students try to interpret a real event that happened to a fleet in the Atlantic Ocean when, after a storm, compasses became disoriented. The interpretation and the creation of an imaginary dialogue between the captains of the ships with Øersted allow students to accept the relationship discovered by the scientist and use it in everyday life. This helps students to interpret phenomena scientifically and evaluate the importance of scientific thinking and experimentation in the development of science.

This activity is the basic unit of planning and analysis. The higher level of collective activity is directed towards the goal of understanding the relationship of electricity to magnetism.

Fifth activity – From electricity to magnetism – electromagnet. The discovery of the relationship between electricity and magnetism by Øersted triggered significant inventions that laid the foundations of modern civilization. One of these is the invention of the electromagnet. The main part of this activity is for students to realize the various applications related to the discovery of the relationship between magnetic and electrical phenomena, especially the electromagnet.

This activity is addressed in the classroom and the objective is mediated by historical evidence of the efforts of scientists to apply in practice the relationship between electricity and magnetism and the construction of an electromagnet on the computer first and then with real materials.

Table 7. The design of the fifth activity according to the expansive cycle.

Questioning	Introduction of the topic under discussion. Students are familiar with the efforts of scientists in the early 19th century to create practical applications that demonstrate the relationship of electricity and magnetism.
Analyzing the situation	Students experiment by constructing a coil to detect the presence of magnetic field.
Modeling	Students construct the model of an electromagnet on the computer screen and experiment with it.
Examining the new model	Students construct an electromagnet with real materials and test it on computer.
Implementing the model	Students apply the operation principle of the electromagnet's construction. They are familiar with its operation and they describe it.
Reflecting and evaluating	Students find and discuss various applications of the electromagnet in everyday life.

The activity begins with a vignette from the history of science that shows the continued effort of scientists in the early 19th century to find practical applications for connecting electricity with magnetism. Particularly highlighted are the efforts of André Marie Ampère and the construction of the coil and magnetic field. (It is worth noting that there is no reference for the students about the meaning of the field, something they perceive only intuitively.) In the first stage of the activity, the situation is analyzed, creating an incentive. Students are familiar with the coil and create its model on the computer. Then they construct a coil and find out through experimentation that the coil behaves as a magnet and is even more powerful than a simple current-carrying conductor. They place a core in the coil and find out that the attractive force of construction is multiplied. In this way, they construct a magnet and manage to control its action as long as they wish while the pulling power is not present when the coil does not have an electricity flow.

In the next stage of integration of the model the students are asked to describe the operation of the electric bell. This device is in their school, and they experience its operation on a daily basis. Seeing the model of the bell on the computer or in the classroom and a real bell enables them to verify the implementation of the electromagnet in devices of everyday life and recognize the value of the construction of the electromagnet. In the process of accepting and evaluating, students are asked to find other devices that make use of the electromagnet. They evaluate its importance in pictures of devices and find other examples.

Sixth activity – New discoveries that changed the world – the motor. The relationship between electricity and magnetism was completed by Faraday, whose contribution was catalytic. In the next two activities, students complete their study on the relationship of power with magnetic phenomena and their interactions.

In the first activity the students/subjects are familiar with the working principle of the electric motor and construct their own motor with simple materials; this project is the subject of activity mediated by cultural tools of the computer and Faraday's efforts to construct the electric motor. In this activity, the objective is mediated by a corresponding software through which students become aware of Faraday's efforts to construct a model of an electric motor. The rules and restrictions imposed by operation of the class, and of course the participation of each student or group of students in conducting activities complement the rest of the elements of the activity.

The structure of the activity follows the design of the expansive cycle of Engeström, whereby the first stage is a simple power function, creating an incentive, while in the second stage the current situation is analyzed: Scientists have reached an impasse regarding the interpretation of the relationship between electricity and magnetism, and Faraday's concerns bring new challenges to the scientific community.

146

Table 8. The design of the sixth activity according to the expansive cycle.

Questioning	Introduction of the topic under discussion. Students are familiar with a leading scientist, Michael Faraday, and the problem he was asked to account for in the interpretation of electromagnetism.
Analyzing the situation	Students summarize Faraday's concerns and the situation that existed during the second and third decades of the 19th century in the field of electromagnetism.
Modeling	Students follow the model of Faraday's experiment on the computer, acknowledge its parts, and monitor its operation.
Examining the new model	The students put the model into operation. They try to construct it with simple materials and describe its function.
Implementing the model	Students using the construction of Faraday's motor give their own interpretation and describe its operation principle.
Reflecting and evaluating	Students find and discuss various applications of the electric motor in everyday life.

In the third stage of the mediation of the modeling tool computer, students construct a virtual model of Faraday's motor and put it into operation. The electric motor is a fact, and students are invited to proceed to the next stage to accept and integrate Faraday's model, which they have created with simple materials (a magnet, a battery, and a coil) and thus produced a prototype motor.

The evaluation and acceptance of the model is amplified in the next stage of activity, in which students give their own interpretations and describe the operation of the motor. Finally they find devices from everyday life and discuss the motor's uses in these devices.

Seventh activity – New discoveries that changed the world – the generator. One question which troubled Faraday remained to be answered to complete the connection of electricity with magnetism: Once a pipeline that is powered by electric current behaves like a magnet, what would happen if a magnet moved in a circular pipe?

In this activity the students/subjects answer Faraday's question by integrating the relationship of electricity with magnetism with an electric current by moving a magnet inside a coil, which is the subject of the activity with the mediation of the computer's cultural tools and the construction of the first single generator by Faraday.

In this activity, the objective is mediated by the software we developed, in which students experiment with the model of Faraday's electrical generator. The rules and restrictions imposed by the operation of the class and of course the participation of each student or group of students in conducting activities complement the rest of the elements of the activity.

Table 9. The design of the seventh activity according to the expansive cycle

Questioning	Students are puzzled by the question that troubled Faraday.
Analyzing the situation	Students analyze the situation as it has been formed with the construction of the electromagnet and the motor.
Modeling	Students construct a virtual model of an electric generator on the computer screen.
Examining the new model	Students put the constructed model into operation.
Implementing the model	Students describe the creation of electric power in factories using the electric generator.
Reflecting and evaluating	Students express their views about the importance of Faraday's work and about the electricity as a foundation of modern civilization.

The structure of the activity follows the design of the expansive cycle of Engeström, whereby in the first and second stages, students are concerned with the question which troubled Faraday and analyze the situation as it had been introduced thus far: The magnetic needle deviates when located near a current-carrying conductor, the attraction takes place when the conductor is circular (coil), and results in the construction of the electromagnet by Henry and of the electric motor by Faraday.

In the third stage of the modeling tool mediation computer, students construct a virtual model of the generator and put it into operation. This results in the production of electricity. The construction of electric power is a fact and students are invited to the next stage to accept and integrate Faraday's model through their knowledge that the electricity reaching their homes is created in this way in power stations.

The evaluation and acceptance of the model is amplified in the next stage of activity, in which students evaluate the importance of electricity generation in modern culture.

Discussion and Further Research

In order to evaluate the design of the teaching package, we implemented it in a primary school class. Moving to the analysis of a complete activity, the mediation offered by the system can be studied based on three levels: epistemological, methodological, and social interaction (reciprocal help) (Bottino et al., 1999). Each of these three levels expresses the role of mediation in relation to each of the components of the mediation activity (tools, rules, division of labor) in the reference model activity.

1. *Epistemological level.* This level deals with the historical and cultural development of the object (in our case the goal of teaching) and the contradictions that characterize this evolution. The analysis at this level takes into account the characteristics of the used tools that are ICT and the history and philosophy of science, incorporating a specific culture that affects the activity itself. That is, how can the tool support learning about the subject of activity and how can it enable the automation of new working methods and construct new tools for activities of real life?

- Students worked in a context in which the scientists dispute about the nature of electricity emerged.
- They conducted substantial activities on the computer to understand the creation of direct current power.
- They conducted activities with materials that verified the model by creating power introduced by the computer.
- They developed skills for applying and comparing the model to real life situations.

2. *Methodological level.* At this level, the analysis is related to the actions and targets involved in the activity mediated by tools. Here it is necessary to understand how the use of tools can support the acquisition of a methodology for achieving the objective of the activity. The support tool includes the construction of new modes of communication structure in the activity.

The tools used by students (the computer and the history and philosophy of science) made it possible for students to:

- Manipulate objects in a context where they could give meaning to problem-solving activities.
- Convert objects that represent abstract concepts with the help of feedback.
- Connect, with a suitable connection, the elements with specific actions for handling objects.
- Create appropriate and useful communication actions to resolve the contradictions that have emerged throughout the learning situation.

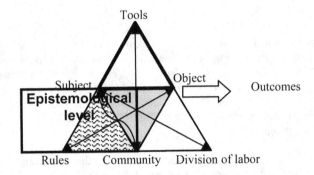

Figure 6. The analysis of activity with emphasis on the epistemological level.

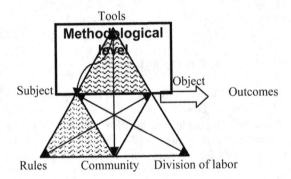

Figure 7. The analysis of activity with emphasis on the methodological level.

3. *Level of social interaction.* The changes in the structure of social relations are considered along with new roles introduced by the mediation tools. The importance of these changes is taken into account in order to create new forms of assistance that might better meet the needs of students. The important issue on this level is the support of students within the zone of forthcoming development as a key link in the development of learning (Vygotsky, 1978).

– The students posed questions that led to the acquisition of skills for designing a solution strategy.
– They offered examples of an effective course of action.
– They generalized.

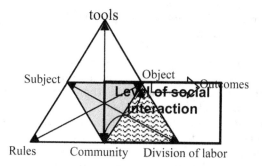

Figure 8. The analysis of activity with emphasis on the level of social interaction

CONCLUSIONS

In our project we propose a methodology for the design and analysis of activities in science teaching in primary education that relies on a corresponding proposal for teaching mathematics (Bottino et al., 1999) that is based on activity theory and expansive learning as formulated and implemented by Engeström.

Our methodology includes design activities within the stages of an expansive cycle (question, analysis, modeling, application of the model, evaluation, and acceptance). For each stage, an appropriate action was designed for students to help them internalize the concept/purpose of the activity.

In the first activity the goal/object was the separation of magnetic and electric phenomena, as introduced by Gilbert. In this activity the students worked on the computer and then used real materials to formulate a scientific method of experimentation such as that first introduced by Gilbert. In the second activity students discover electrostatic attraction. We used vignettes of historic figures from the work of Gilbert, du Fay, and Franklin. In the third activity the goal was the construction and operation of the battery. The confrontation between Volta and Galvani was the basic idea around which the students worked to understand the operation and construction of the battery. In the fourth activity, the students/ subjects examine Øersted's model experiment on the computer and then perform the experiment themselves with real materials. In the fifth activity, students construct an electromagnet. In the last two activities, which are inspired by the life and work of Faraday, students use software and elements from the history and philosophy of science to create an electric generator and an electric motor.

The methodology also includes the analysis of the components of the activity (subject, object, community), their mutual relations, and entities that mediate these relationships (tools, rules, division of labor). In such a context we analyze the activity from three perspectives: epistemological-methodological, social interaction, and mutual aid. These three perspectives are respectively related to the relations of subject-community, subject-object, and object-community, which are affected by the mediation resulting from the use of new technologies and data from the history and philosophy of science.

The students worked in a context that includes many years of effort spent by scientists to prove the unity of electric and magnetic phenomena; they performed activities on the computer, and using real materials, they handled objects and created relevant and useful communication actions that resolved the contradictions, raised questions, and made generalizations.

The results of the case study that we examined showed that the mediation of the tools used (software, data from the history and philosophy of science) and the examples offered to help students, played a key role in the activity's success. The students provided a solution to the conflict over what creates electricity, and concluded the structure of the concept.

At this point I agree with Yves Clot (2009, p. 302), who states that three results have been obtained from Engeström's contribution to the development of intervention studies in workplaces and curriculum in science education. The first result stresses that action for transforming work is the condition of the production of scientific knowledge. The second result attests to the importance of the collective in the development of activity. The third result concerns the question of models in the intervention. The development of the scientific concepts of the interventionist and the spontaneous concepts in the action of the professionals is accomplished along lines that cross but do not become identical.

The encouraging results of our efforts point to the need for further investigation of this methodology in more activities, and we intend to initiate such investigations in the near future.

NOTES

[1] Yrjö Engeström: Learning by expanding: Ten Years after, Introduction to the German edition of *Learning by Expanding*, published in 1999 under the title *Lernen durch Expansion* (Marburg: BdWi-Verlag; translated by Falk Seeger); also in the Japanese edition, published in 1999 under the title *Kakucho ni yoru Gakushu* (Tokyo: Shin-yo-sha; translated by a group led by Katsuhiro Yamazumi). http://lchc.ucsd.edu/MCA/ Paper/Engestrom/expanding/intro.htm.

[2] Greek philosopher Thales of Miletus (~624–547 BC).

[3] It was Plato (427–347 BC) in the 4th-century BC who first made a reference that has survived to the present day on "... that marvellous attraction exercised by amber and by the lodestone ..." in one of his dialogues, the Timaeus.

[4] On this point see Dagenai (2003): The following scientists-Aristotle (384 BC–322 BC), Copernicus (1473–1543), Galileo (1564–1642), Newton (1650–1727), Watt (1736–1819), Young (1773–1829) and Bohr (1885–1962) can form the basis for creating a story line. In addition, Gilbert (1544–1603), DuFay (1698–1739), Franklin (1706–1790), Galvani (1737–1798), Volta (1745–1827), Øersted (1777–1851), Ampère (1775–1835), and Faraday (1791–1867) formed the basis for creating story lines in the context of Science Teacher e-Training (STeT) program: Teaching Science using case studies from the History of EU Science (Comenius 2.1) http://valanides.org/ScienceTeaching/Lessonplans/tabid/78/Default.aspx .

REFERENCES

Allchin, D. (1997). Rekindling phlogiston: From classroom case study to interdisciplinary relationships. *Science & Education, 6,* 473–509.

Barab, S. A., Cherkes-Julkowski, M., Swenson, R., Garrett. S., Shaw, R. E., & Young, M. (1999). Principles of self-organization: Ecologizing the learner-facilitator system. *Journal of the Learning Sciences, 8*, 349–390.

Barab, S. A., Evans, M., & Baek, E.-O. (2003). Activity theory as a lens for charactering the participatory unit. In D. Jonassen (Ed.), *International handbook on communication technologies* (Vol. 2, pp. 199–214). Mahwah, NJ: Lawrence Erlbaum Associates.

Barab, S., Schatz, S., & Scheckler, R. (2004). Using activity theory to conceptualize online community and using online community to conceptualize activity theory. *Mind, Culture, and Activity, 11*(1), 25–47.

Basharina, O. (2007). An activity theory perspective on student – Reported contradictions in international telecollaboration. *Language Learning & Technology, 11*(2), 36–58.

Bevilacqua, F., & Giannetto, E. (1998). The history of physics and European physics education. In *International handbook of science education* (Vol. II, pp. 1015–1026). Dordrecht: Kluwer Academic Publisher.

Bigum, C. (2000). Actor-network theory and online university teaching: Translation versus diffusion. In B. A. Knight & L. Rowan (Eds.), *Researching futures oriented pedagogies* (pp. 7–22). Flaxton, Qld: PostPressed.

Binnie, A. (2001). Using the history of electricity and magnetism to enhance teaching. *Science & Education, 10*, 379–389.

Bottino, R-M., Chiappini, G., Forcheri, P., Lemut, E., & Molfino, M-T., (1999). Activity theory: A framework for design and reporting on research projects based on ICT. *Education and Information Technologies, 4*(3), 281–295.

Bruner, R. (1975). From communication to language – A psychological perspective. *Cognition, 3*, 255–287.

Butterfield, H. (1994). *The origins of modern science*. London: G. Bell & Sons (Greek edition).

Cole, M. (1988). Cross-cultural research in the sociohistorical tradition. *Human Development, 31*, 137–151.

Cole, M. (1996). *Cultural psychology. A once and future discipline*. Cambridge, MA: Harvard University Press.

Cole, M. (1999). Cultural psychology: Some general principles and a concrete example. In Y. Engestrom, R. Miettinen, & R. Punamaki (Eds.), *Perspectives on activity theory*. New York: Cambridge University Press.

Cole, M., & Engeström, Y. (1993). A cultural-historical interpretation of distributed cognition. In G. Salomon (Ed.), *Distribute cognition: Psychological and educational considerations*. Cambridge: Cambridge University Press.

Clot, Y. (2009). Clinic of activity theory: The dialogue as instrument. In A. Sannino, H. Daniels, & K. Gutierrez (Eds.), Learning and expanding with activity theory. Cambridge: Cambidge University Press.

Dafermos, M. (2002*). The cultural-historical theory of L. S. Vygotsky*. Athens: Atrapos [in Greek].

Dagenais, A., (2003). Teaching the history of science without lectures. In *Proceedings of 7th International History, Philosophy of Science and Science Teaching Conference* (pp. 227–234). Winnipeg.

Dagenais, A., (2010). Teaching high school physics with a story-line. *Interchange, 41*(4), 335–345.

Davydov, V. (1999). The content and unsolved problems of activity theory. In Y. Engeström, R. Miettinen, & R. Punamaki (Eds.), *Perspectives on activity theory*. New York: Cambridge University Press.

Dodge, B. J. (2001). Focus five rules for writing great webquests. *Learning and Leading with Technology, 28*(8), 6–9.

Duignan, M., Noble, T., & Biddle, R., (2006). Activity theory for design from checklist to interview. In T. Clemmensen, P. Campos, R. Omgreen, Al. Petjersen, & W. Wong (Eds.), *IFIP International federation for information processing* (Vol. 221, Human Work Interaction Design: Designing for Human Work, pp. 1–25). Boston: Springer.

Elmore, F. R. (2004). Foreword. In E. Coppola (Ed.), *Powering up: Learning to teach well with technology*. New York: Teachers College Press.

Engeström, Y. (1987). *Learning by expanding: An activity-theoretical approach to developmental research*. Helsinki: Orienta-Konsultit.

Engeström, Y. (1999a). Activity theory and individual and social transformation. In Y. Engeström, R. Miettinen, & R. Punamaki (Eds.), *Perspectives on activity theory*. New York: Cambridge University Press.

Engeström, Y. (1999b). Innovative learning in work teams: Analyzing cycles of knowledge creation in practice. In Y. Engeström, R. Miettinen, & R. Punamaki (Eds.), *Perspectives on activity theory*. New York: Cambridge University Press.

Engeström, Y. (1999c). Learning by expanding: Ten years after. Introduction to the German edition of *Learning by expanding*, published in 1999 under the title *Lernen durch Expansion* (Marburg: BdWi-Verlag; translated by Falk Seeger). Retrieved November 9, 2010, from http://lchc.ucsd.edu/MCA/Paper/Engestrom/expanding/intro.htm

Engeström, Y. (2001). Expansive learning at work: Toward an activity theoretical reconceptualization. *Journal of Education and Work, 14*(1), DOI: 10.1080/13639080020028747.

Engeström, Y., & Miettinen, R., (1999). Introduction. In Y. Engeström, R. Miettinen, & R. Punamaki (Eds.), *Perspectives on activity theory*. New York: Cambridge University Press.

Engeström, Y., Engeström, R., & Vahaaho, T., (1999). When the center does not hold: The importance of knotworking. In S. Chaiklin, M. Hedegaard, & U. J. Jensen (Eds.), *Activity theory and social practice: Cultural-historical approaches*. Aarhus: Aarhus University Press.

Faraday, M. (1835–1855). *Experimental researches in electricity* (reprinted). New York: Dover.

Galili, I., & Hazan, A. (2000). The influence of an historically oriented course on students content knowledge in optics evaluated by means of facets-schemes analysis. *Physics Education Research: A Supplement to the American Journal of Physics, 68*(7), S3–S15.

Gillispie, C. C. (1994). *The edge of objectivity: An essay in the history of scientific ideas*. Princeton, NJ: Princeton University Press, 1960 [in Greek].

Griffin, P., & Cole, M. (1984). Current activity for the future: The zoped. In B. Rogoff & J. V.Wertsch (Eds.), *Children's learning in the zone of proximal development*. San Francisco: Jossey-Bass.

Guisasola, J., Almudí, J. M., & Furió, C. (2005). The nature of science and its implications for physics textbooks. *Science & Education, 14*(3), 321–328.

Hakkarainen, P., Engeström, R., Kangas, K., Bollström-Huttunen, M., & Hakkarainen, K. (2004). The artefact project – Hybrid knowledge building in a networked learning environment. Paper presented at the Scandinavian Summer Cruise at Baltic Sea, "Motivation, learning, and knowledge building in the 21st century," , June 18–June 21.

Heering, P. (2003). History-science-epistemology: On the use of historical experiments in physics teacher training. In W. F. McComas (Ed.), *Proceedings of the 6th IHPST Conference*, Denver 2001. Avaible from the IHPST Group, IHPST.org.

Heering, P. (2005). Analysing unsuccessful experiments and instruments with the replication method. In *ÉNDOXA: Series Filosóficas* (Vol. 19, pp. 315–340). Madrid: UNED.

Heilbron, J. L. (1979). *Electricity in the 17th and 18th centuries: A study of early Modern physics*. Berkeley, CA: University of California Press.

Heisenberg, W. (1997). *Das Naturbild der heutigen Physik*. Rowohlt, 1955 [in Greek].

Henke, A., Höttecke, D., & Riess, F. (2009). Case studies for teaching and learning with history and philosophy of science exemplary results of the HIPST project in Germany. Paper presented at the Tenth International History, Philosophy, and Science, Teaching Conference University of Notre Dame South Bend, USA, June 24–28.

Irwin, A. R. (2000). Historical case studies: Teaching the nature of science in context. *Science Education, 84*(1), 5–26.

Jonassen, D. (2000). *Computers as mindtools for school: Engaging critical thinking*. Upper Saddle River, NJ: Merrill.

154

Jonassen, D., & Rohrer-Murphy, L. (1999). Activity theory as a framework for designing constructivist learning environments. *Educational Technology Research and Development, 47*(1), 61–79.

Kaptelinin, V., & Nardi, B. (2006). *Acting with technology: Activity theory and interaction design.* Cambridge: The MIT Press.

Kaptelinin, V., Nardi, B. A., & Macaulay, C. (1999). The activity checklist: A tool for representing the "Space" of context. *ACM / Interactions, Methods & Tools, 6,* 27–39.

Kipnis, N. (2005). Chance in science: The discovery of electromagnetism by H.C. Oersted. *Science Education, 14,* 1–28.

Koschmann, T. (1996). Paradigm shifts and instructional technology. In T. Koschmann (Ed.), *CSCL: Theory and practice of an emerging paradigm.* Mahwah, NJ: Lawrence Erlbaum.

Kozma, R. (2003). *Technology, innovation, and educational change: A global perspective.* Eugene, OR: International Society for Educational Technology.

Kutti, K. (1996). Activity theory as a potential framework for human-computer interaction research. In B. Nardi (Ed.), *Context and consciousness.* London: MIT Press.

Latour, B. (1988). *The pasteurisation of France.* Cambridge: Harvard University Press.

Lee, C. D., & Smagorinsky, P. (2000*). Vygotskian perspectives on literacy research, constructing meaning through collaborative inquiry.* Cambridge: Cambridge University Press.

Leontiev, A. (1979). The problem of activity in psychology. In J. Wertsch (Ed.), *The concept of activity in soviet psychology.* New York: Armonk, M.E. Sharpe.

Lim, P. C. (2007). Effective integration of ICT in Singapore schools: Pedagogical and policy implications. *Education Tech Research, 55,* 83–116.

Luria, A. (1976). *Cognitive development: Its cultural and social foundations.* Cambridge: Harvard University Press.

Malamitsa, K., Kokkotas, P., & Stamoulis, E. (2005). *The use of aspects of history of science in teaching science enhances the development of critical thinking.* Paper presented at the Eighth International History, Philosophy and Science Teaching Conference (IHPST 8), Teaching and Communicating Science: What the history, philosophy and sociology of science can contribute, England, 15–18 July 2005, Abstracts (pp. 63–64), Leeds: University of Leeds. Retrieved from http://www.ihpst2005. leeds.ac.uk/papers.htm.

Masson, S., & Vázquez-Abad, J. (2006). Integrating history of science in science education through historical microworld to promote conceptual change. *Journal of Science Education and Technology, 15*(3), 257–268.

Matthews, M. (1994). *Science teaching, The role of history and philosophy of science.* New York: Routledge.

Matthews, M. (1998). The nature of science and science teaching. In B. Fraser & K. Tobin (Eds.), *International handbook of science education* (Pt. 2). Dordrecht/Boston/London: Kluwer Academic Publishers.

Monk, M., & Osborn, J. (1997). Placing the history and philosophy of science on the curriculum: A model for the development of pedagogy. *Science Education, 81,* 405–424.

Mwanza, D. (2000). *Mind the gap: Activity theory and design.* Paper submitted at CSCW 2000 Conference in Philadelphia, PA, December 2–6.

Nardi, B. A. (1996). Activity theory and human-computer interaction. In B. A. Nardi (Ed.), *Context and consciousness: Activity theory and human-computer interaction* (pp. 69–103). Cambridge and London: MIT Press.

Nersessian, N. J. (1995). Should physicists preach what they practice? Constructive modeling in doing and learning physics. *Science & Education, 4*(3), 203–226.

Nersessian, N. J. (2002). Abstraction via generic modeling in concept formation in science. *Mind & Society, 5*(3), 129–154.

Nersessian, N. J. (2009). Conceptual change: Creativity, cognition, and culture. In J. Meheus & T. Nickles (Eds.), *Models of discovery and creativity* (pp. 127–158). New York: Springer.

Ravanis, K. (1999). *Science in preschool education: A teaching and cognitive approach.* Athens: Typothito [in Greek].

Rizzo, A. (2003). Activity Centered Professional Development and Teachers. Take-Up of ICT, paper was presented at the IFIP Working Groups 3.1 and 3.3 Working Conference: ICT and the Teacher of the Future, held at St. Hilda's College, The University of Melbourne, Australia, 27–31 January.

Rossi, P. (2004). *La nascita della scienza in Europa*. Greek translation by Tsiamouras Panagiotis. Athens: Ellinika Grammata.

Roth, W. M. (2009). On the inclusion of emotions, identity and ethico-moral dimensions of actions. In A. Sannino, H. Daniels, & K. Gutierrez (Eds.), *Learning and expanding with activity theory*. Cambridge: Cambidge University Press.

Roth, W.-M., & Lee, S. (2004). Science education as/for participation in the community. *Science Education, 88*(2), 263–291.

Roth, W.-M., & Lee, Y. J. (2007). Vygotsky's neglected legacy: Cultural-historical activity theory. *Review of Educational Research, 77*(2), 186–232.

Segrè, E. (2001*). From falling bodies to radio waves: Classical physicists and their discoveries*. Greek translation by Kostantina Mergia, Vol. A, Athens: Diavlos.

Seker, H., & Welsh, L. C. (2006). The use of history of mechanics in teaching motion and force units. *Science & Education, 15*, 55–89.

Seroglou, F., & Koumaras, P. (2001). The contribution of the history of physics in physics education: A review. *Science & Education, 10*, 153–172.

Seroglou, F., Koumaras, P., & Tselfes, V. (1998). History of science and instructional design: The case of electromagnetism. *Science and Education, 7*, 261–280.

Sneider, C. I., & Ohadi, M. M. (1998). Unraveling students' misconceptions about the earth's shape and gravity. *Science Education, 82*, 265–284.

Stamoulis, E., & Kokkotas, P. (2006). Using activity theory to analyze the effect of input from HPS in a technologically rich environment for teaching science. In *Proceedings of 3rd National Conference Early Childhood Education of University Thessaly on Science Education: Learning Methods and Technologies*, Volos.

Stamoulis, E., Kokkotas, P., & Mavrogiannakis, M. (2003). The contribution of history and philosophy of science in their teaching: Presentation of Archimedes and his work with the software. In *Proceedings of the 2nd National Conference: The contribution of history and philosophy of science in science teaching*, May 8–11 (pp. 468–473). Athens: University of Athens.

Stinner, A., & Williams, H. (1998). History and philosophy of science in the science curriculum. In B. Fraser & K. Tobin (Eds.), *International handbook of science education* (Pt. 2). Dordrecht/Boston/ London: Kluwer Academic Publishers.

Stinner, A. (1994). The story of force: From Aristotle to Einstein. *Physics Education, 29*(2), 77–85.

Stinner, A. (1995). Contextual settings, science stories, and large context problems: Toward a more humanistic science education. *Science Education, 79*(5), 555–581.

Stinner, A., MacMillan, B., Metz, D., Jilek, J., & Klassen, S. (2003). The renewal of case studies in science education. *Science & Education, 12*, 617–643.

Stinner, A. (2007). Toward a humanistic science education: Using stories, drama, and the theatre. *Canadian Theater Review*, 14-19.

Thorne, S. (2003). Artefacts and cultures-of-use in intercultural communication. *Language Learning & Technology, 7*(2), 38–67.

Voutsina, L., & Ravanis, K. (2007). Historical models and mental representations of students' school on magnetism. In D. Koliopoulos (Ed.), *History, philosophy and teaching science*. Maroussi: Othisi [in Greek].

Vygotsky, L. (1978). *Mind in society: The development of higher psychological processes*. Cambridge: Harvard University Press.

Wandersee, J. (1985). Can the history of science help science educators anticipate students' misconceptions? *Journal of Research in Science Teaching, 23*, 581–597.

Wartofsky, M. (1979). *Models. Representation and the scientific understanding*. Boston: Reidel.

Wells, G. (2000). From action to writing: Modes of representing and knowing. In J. W. Astington (Ed.), *Minds in the making*. Oxford: Blackwell Publishers.

Wertsch, J. V. (1985). *Vygotsky and the social formation of mind*. Cambridge: Harvard University Press.

Wertsch, J. V. (1991). *Voices of the mind: A sociocultural approach to mediated action*. Cambridge: Harvard University Press.

Westfall, R. S. (2006). *The construction of modern science. Mechanism and mechanics*. New York: Cambridge University Press (in Greek language edition).

Whittaker, E. T. (1987). *A history of the theories of aether and electricity from the age of Descartes: The close of the nineteenth century*. Dublin: Dublin University press.

Woolgar, S. (2003*). Science: The very idea*. London: Routledge (in Greek language, Katoptro, Athens).

Yoon, H.-G. (2006). *The nature of science drama in science education*. Paper presented at the 9th International Conference on Public Communication of Science and Technology, Coex, Seoul, Korea, May 17–20. Retrieved November, 2010, from http://sciencedrama.cnue.ac.kr/admin/upload/non/yoon(2006).pdf.

Efthymis Stamoulis
School of Education
University of Ioannina
Greece

Katerina Plakitsi
School of Education
University of Ioannina
Greece

7. UNIVERSITY SCIENCE TEACHING PROGRAMS[1]

What's New in Lab Activities from a Chat Context?
The Case of Magnetism

INTRODUCTION

This is a research study on connecting science education with Cultural- Historical Activity Theory (CHAT). It focuses on a sequence of science education lab activities on the topic of magnetism that was organized at the University of Ioannina in Greece. When discussing activity, activity theorists are not simply concerned with "doing" as a disembodied action but are referring to "doing in order to transform something," with the focus on the contextualized activity of the system as a whole (Engeström, 1987; Kuutti, 1996). We focus on object, which is constantly in transition and under construction, and "it manifests itself in different forms for different participants and at different moments of the activity" (Engeström, 1999). The Laboratory Lesson of Magnetism based on Activity Theory (LLMAT) is an experimental course at the University of Ioannina (Theodoraki & Plakitsi 2009). The research is divided into seven steps: organization of the LLMAT, video studies, evaluation, questionnaires and interviews, elaboration of data, students' practice, and reflection. The functions of objects are interpreted as well as other connected factors such as are subjects, tools, rules, community, and division of labor. This study, even though limited, could push forward the boundaries of teachers' training in science education. It especially highlights teachers' capability to provoke meaningful learning in school science classrooms.

Moreover this study approaches the problem of inefficiency in science education, a problem identified by the Programme for International Student Assessment (PISA) 2009. PISA underlines, the need for many advanced countries to tackle educational underperformance so that as many members of their future workforces as possible are equipped with at least the baseline competencies that enable them to participate in social and economic development. This program also presents the results of teaching science education in various countries.

The research is an innovation which follows the shift in sociocultural education in science education and is supported by scientists worldwide (Cole, 2006; Engeström, 2005; Plakitsi, 2008; Roth & Lee, 2004). Cultural-Historical Activity Theory provides a new and more fruitful framework for analyzing and designing science education activities, with the aim of achieving scientific literacy in the early years (5–9 years old).

K. Plakitsi (ed.), Activity Theory in Formal and Informal Science Education, 159–195.

The methodological framework of this study is inspired by the developmental work research of Engeström (2005), the design experiment method (Roth, 2007), and the project "The 5th Dimension" (Cole, 2006). The study contributes to the dissemination of cultural- historical activity theory in science education in Europe. This approach contributes to rethinking of scientific literacy (Roth & Lee, 2004), and the role of information and communication technologies (Kaptelinin & Nardi, 2006; van Eijck & Roth, 2007).

Activity theory is a psychological and multidisciplinary theory with a naturalistic emphasis that offers a framework for describing activity and provides a set of perspectives on practice that interlink individual and social levels (Engeström, 1987; Nardi, 1996; Roth & Lee, 2004). Activity theory is a theoretical framework for analyzing human practices as developmental processes with individual and social levels simultaneously interlinked (Kaptelinin & Nardi, 2006; Kuutti, 1996; Nardi, 1996). This framework uses "activity" as the basic unit for studying human practices.

The relations between participant and object are not direct; rather, they are mediated by various factors, including tools, community, rules, and division of labor. From an analytical point of view it is quite complicated to handle this. To overcome this obstacle we adopted Engeström's triangular model (1987).

Our research, concentrates on transferring the activity theory into the field of science education. As it is combined with other relevant case studies it finally aims to validate the activity theory as an evaluation tool of scientific activities in different learning environments, as it is a school classroom, a laboratory etc.

The crucial role of this study is to define:
- What kind of tools do the subjects use to achieve their objectives and how?
- What kind of rules affect the way the subjects achieve the objectives and how?
- How does the division of labor influence the way the subjects satisfy their objectives?
- How do the tools in use affect the way the community achieves the objectives?
- What rules affect the way the community satisfies their objectives and how?
- How does the division of labor affect the way the community achieves the objectives?

RATIONALE

In this paper, we conduct pilot research to test Activity Theory as an analytical tool for science education activities. Activity theory is a theory with expanding applications in different fields of science as well as in natural sciences education. The unit of analysis is the activity. Students work in groups defined in the community, and they use intermediary tools towards a common goal. The objects play an important role in establishing new principles or ideas in the context of rules that the entire community follows. This provides flexibility in moving from one activity to another and takes advantage of previous knowledge. Thus, the construction of knowledge becomes meaningful for students who interact with one another, as well as with tools and means within the community of learners and within the activity's context (Engeström, 1999). Within this framework, we first present a succession of laboratory activities concerning the magnetic properties of certain materials. Then we analyze the framework of activity theory. The discussion concerns some limits of the transfer of activity theory to the science education context (Cole, 1995). The research is based on a case study and uses multiple methods for gathering data (Yin, 1994). We construct the LLMAT lab lesson (Laboratory Lesson of Magnetism based on Activity Theory), which is an experimental course in the Department of Early Childhood Education in the field of science education. We use this lab lesson as a methodological tool to prepare university students to teach the magnetism issue in the classroom. The central point of LLMAT is our position about educational success or failure, which has been explicated as a collective activity in the social framework. The organization of LLMAT included the following steps:

1. Awareness. In this stage we codecided that the topic of magnetism was the most interesting topic for our studies.

2. Comparison. We compared the laboratory lesson of magnetism with previous laboratory lessons, and we used obtained knowledge, skills, and materials.

3. Exploration/Activating Prior Learning. In this stage we tried to trace some students' obstacles in their conceptualization of magnetism.

4. Creation. We developed additional insights about magnetism, and we created activities based on our students' needs.

Students use their prior knowledge of this issue to categorize some materials by testing their behavior when they come close to a magnet. Then they obtain further knowledge about magnetic properties and create new activities based on their prior knowledge. Furthermore, they take into account pupils' cognitive obstacles in organizing the new activities so as to achieve the aim.

In recent decades, many researchers in the United States, Canada, and Europe have developed a new theoretical foundation and research methods on the Activity Theory, and crucial issues such as "learning communities," the "motivation to learn," and "quality of natural science research" are the most important topics in recent European research magazines, conferences, and books.

Studying the results of PISA 2009 highlights the need for a new more fruitful framework to achieve the goal of scientific literacy and science education. This could be achieved by using Cultural- Historical Activity Theory (CHAT). We stress the importance of the possibility that the theory could bridge the gap between theory and praxis.

It can also lead to the goal of interdisciplinary education in science education in the context of multicultural Europe. Therefore, a new mentality is emerging in science education as a social process. This could reshape education in natural sciences from the inside in a natural and logical manner in the context of lifelong learning.

The theoretical and methodological framework of analysis and design activities is the developmental approach to research of Yrjö Engeström (2005). Before the introduction of CHAT in the laboratory lessons, the students in most cases were usually treated as passive recipients of information, whereas the teachers took on the role of an information dispenser, authoritative expert, and fountainhead of information and knowledge. The key elements of our method include the method of Mwanza (Mwanza & Engeström, 2003). The method of Mwanza (2001) consists of eight stages (Eight-step Model of Mwanza):

1 Implementation of activities of interest to children (activity of interest). We focus on children's interests to determine which activities we carry out.
2. The aim of the activity (objective of activity). We find the aim–object of the activities that we carry out.
3. Subject activity (subject in this activity). We find the different subjects who act during the activity (teacher, students, etc.)
4. Tools of the activity (tools mediating activity). We explore the different tools that subjects decide to use during the activity.
5. Mediators to regulate the activity (rules and regulations mediating the activity). We define the rules that subjects decide to use during the activity.
6. Division of labor mediates the activity (division of labor mediating the activity). We focus on the way that subjects decide to divide their work in small groups during the activity.
7. Community hosting the community in which activity is conducted. We define the community in which the activity is taking place.
8. Results (outcomes). We define the results-outcomes that result from the activity and create new activities based on the results.

The steps of this methodology are based on the expanding learning of Engeström (1987) and on the strong research tradition of Design Experiment (Brown, 1992; Roth, 2005).

All the ideas that are expressed within the frame of Activity Theory strengthen and expand their practices using "tools" which are traditionally associated with the sociocultural and psychological aspects of the use of "tool." This approach stresses the importance of the "subject's" cultural behavior in relation to the "tool." Thus, the analysis of activities in this methodology takes into account the use of cultural tools. The lack of a standardized method for applying the activity theory could be attributed to the fact that many basic principles of CHAT are configured for the area of ICT and not for the area of Science Education (Kaptelinin, 1996). With this teaching intervention we try to create a complete model using CHAT, with expanding applications in education as well as in natural sciences education.

Activity Theory provides a comprehensive framework for analyzing and designing activities from the field of science education for compulsory education, so as to achieve the goals of scientific literacy, especially in the early years (5–9) years old). Using CHAT we study in depth and analyze the interactions that occur within learning communities. In particular, we explore the social identities of children and the subjective perceptions of their own activity and their role in it. With the current positions of Activity Theory (Engeström, 1999), we analyze the six basic categorical variables of an activity:

1. The subjects of an activity (e.g., teachers, students)
2. The object-target of the activity

3. The mediation tools (e.g., verbal tools, materials, activities, history of science, computer, animation)
4. The operating rules of learning communities
5. The roles of the members of learning communities
6. The division of labor

METHOD

The study is based on Engeström's theory of expanding learning. The ideas presented in CHAT enhance and extend the practical concerns of tool usage, which are traditionally addressed by linking the design solution to sociocultural and psychological aspects of the tool user. This approach highlights the importance of the tool user's cultural behavior revealed during tool usage. It seems that by analyzing human activity in context, using this framework, the computer tool developer can fully account for the complex and intertwining issues that impact the usefulness of the computer tool through its design. Although the ideas presented in this framework sound like a promising method for providing a much needed common vocabulary for describing human activity, there is no standard method for putting Activity Theory ideas into practice (Nardi, 1996). The lack of a standard method for applying Activity Theory could be attributed to the fact that there are several basic principles of Activity Theory (Kaptelinin, 1996) on which one could base their analysis. In addition, the framework itself is continuously evolving, in the sense that concepts from this framework have been interpreted and applied in various ways in different contexts. As a result, difficulties have come up concerning replicating, comparing, and criticizing the approaches taken to apply Activity Theory.

From a sample of 160 third grade University students, three pairs were assigned to teach the magnetic properties of various materials in three different preprimary classrooms during their two-week practice in schools. In order to do this, they had to follow the steps of the LLMAT lab lesson practiced at University. At this point, we have to mention that the students had had little previous experience in a real classroom situation, and they had to work with 18 to 20 pupils in class. Finally, teachers in the schools were willing to collaborate with the students during the whole procedure. Students participated in the designing of lab activities, following the eight-step model of Mwanza (2001).

1. Activity of interest. In this stage students modified the sort of activity in which they were interested. The study of magnets began with an exploration of magnetic and nonmagnetic properties of certain materials. We provided each pair of students with a horseshoe magnet and encouraged them to go on a "magnetic hunt" with their partner. They explored the room, predicting the behavior of different materials when they came close to a magnet.
2. Objective of activity. Students explained the reason the activity takes place. They shared their findings and made observations while experimenting with the magnets.
3. Subject in this activity. They discussed those who were involved in the activity (students, teachers, parents).

4. Tools mediating activity. Books and other materials were the tools with which the subjects (students) carried out the activity. At this point, students demonstrated a deeper understanding of magnets and magnetic properties, knowledge of the earth's magnetic field and the way a compass works, and usage of magnets in everyday life.

5. Rules and regulations mediating the activity. They collectively accepted the rules that all had to follow during the activity: (a) Each pair of students worked together to explore the strength of the magnet. (b) They recorded their findings in their data sheet. (c) When students completed their experimentation, they discussed their findings in the classroom. (d) The teacher evaluated the students' responses.

6. Division of labor mediating the activity. The teacher (a) demonstrated how to set up the experiment (without actually demonstrating the results; this is for the students to discover). (b) The teacher brought the class together for discussion (What happened? Were our hypotheses correct? What conclusions can we make? Which poles repel or attract?). Students either worked in pairs or experimented with a partner to determine the magnetic and the nonmagnetic properties of materials.

7. Community in which activity is conducted. In this step we defined the environment in which the activity was carried out. More specifically, the environment was a class with Greek pupils.

8. Outcomes. This is the final step in which we provided an estimation of the results from carrying out this activity. The results concerning both the outcome of the activity and the way the activity is carried out.

Research Objectives

1. Applying activities for effective science learning with the aim of scientific literacy.
2. Applying activities in accordance with the current positions of CHAT, using the method of Mwanza (Mwanza, 2001).
3. Analysis of activities of natural sciences with the CHAT tool, namely the model of analysis of Mwanza.

Teaching Objectives According to Greek National Curriculum (Table1)

1. To discover magnetic properties. Discrimination of magnetic and nonmagnetic materials.
2. To observe that like poles repel and unlike poles attract each other.
3. To locate poles on various kinds of magnets.
4. To identify the strongest parts of a magnet.
5. To manage the cognitive obstacles of children in understanding the field of magnets and magnetism.

Table 1. Connection with the Greek National Curriculum (5–6 years old).

Content guiding principles	General goals (knowledge, skills attitudes and values)	Indicative fundamental cross-thematic concepts	Frameworks achieved through CHAT
With appropriate teaching interventions both in the nursery school and the wider environment pupils should be provided many opportunities to discover the basic characteristics of the properties of different materials.	Pupils are encouraged to observe and find the similarities and differences between the materials they use. They are also encouraged to classify materials according to common properties such as color, transparency, behavior in water, etc. (studies of the environment, mathematics, music, art). To describe the changes in certain materials under specific conditions (e.g., when put in water, when diluted, etc.). (Studies of the environment, language, physical education, computers, music.)	Individual-Group Interaction System Classification	– Multiple aspects. – Meetings to solve problem (e.g., evaluation, educational program, vocational education, early intervention). – Open communication both as process and as a result. – Communication networks. – Exchange of knowledge and experience. – The evaluation and tools are directly related to curriculum content.

Data collection for this teaching intervention focused not only on children's performances, but also on children's ideas about magnetism and the difficulties in the children's thinking. Data collection includes:

1. A semistructured interview from (a) the teacher and (b) the children.
2. Videotaped activities carried out in order to collect the qualitative data.
3. Data collection through appropriate worksheets designed to investigate children's ideas about magnetism.

The evaluation focuses on exploring children's ideas about magnetism and the ability of children to undertake and perform an activity on their own during the teaching intervention. The evaluation is based on the current positions of CHAT and specifically on the model of Mwanza (2001) and monitoring of children based on Hedegaard (Hedegaard & Fleer, 2008).

Activities

1. Getting to know about magnets and magnetic and nonmagnetic materials.
Materials: (a) Magnets of various shapes and sizes: horseshoe, rodlike, cylindrical, square, magnetic letter board, refrigerator magnets. (b) Magnetic items (iron or steel): needles, batteries, keys, pins, paper clips, scissors. (c) Nonmagnetic items: copper, wood, glass, plastic, cork, paper, sponge.

Target of activity according to CHAT: (a) Observation of children in relation to materials, or "tools," that they choose to play with. (b) Observation of children in relation to the magnets, or "tools," with which they choose to experiment. (c). Identification of words and phrases that are used by children (language "tools") to describe and explain magnetism. (d) Observation of interaction between "subjects" and "community" during the activity. (e) Identifying the "rules" that are followed in the activity. (f) Observation and recording of the use of "tools" from the "subjects" of the "community." (g) Evaluation of the outcome of the activity ("object").

Description of the activity: We provide the children with magnetic and non-magnetic materials and different kinds of magnets. We let them play with the materials freely. Children make observations about the behavior of magnets and magnetic and nonmagnetic materials. Then we give children the first worksheet (Figure 1) to make their predictions and to verify the information. Children make their predictions gradually, for one material at a time, and write down words and phrases they use to justify their answer. Then they test their predictions.

1° WORKSHEET		
OBJECT	PREDICTION	TESTING PREDICTION

Figure 1. First worksheet.

2. Sort out materials according to their behavior when approaching the magnet. Magnets attract each other and attract other objects as well.
Materials: (a) Rodlike magnets, horseshoe magnets; (b) marble, glass, wood, plastic, metal fasteners; (c) aluminum, steel, bronze, and copper.

Target of activity according to CHAT: (a) Investigation and assessment of the cultural level of children regarding their ability to recognize, operate, and use common and laboratory materials ("tools"). (b) Assessment of communication and interaction between the "subjects" of research and the use of "tools." (c) Identification and registration of objects, or "tools," that children can explore. (d) Observation and recording of the interaction of the child-"subjects" with the

166

"learning community." (e) Registration of "rules" to be followed during the activity. (f) Identifying and recording the outcome of the activity ("object").

Description of the activity: The children use the materials as they wish and divide them into two categories: "magnetic materials" and "nonmagnetic materials" (Table 2). Each team of children selects a symbol for each class and sticks samples of magnetic and nonmagnetic objects on the table. During the activity, children observe that attraction forces are a function of the material and are not related to other properties such as shape.

Table 2. Categories of materials.

Magnetic materials	Nonmagnetic material

Then the children are given the same worksheet (second worksheet), on which they are asked to draw materials of their choice in the appropriate position (Fig. 2). In assessing the outcome of the activity, we focus on the materials the children choose to draw. In this way we see the relationship of different "*tools*" of CHAT with the "*subjects.*" In particular we will see if the children are familiar with the operation and use of magnets and magnetic and nonmagnetic materials. The manner in which children choose the "*tools*" helps us to understand whether the framework of CHAT helps children develop scientific knowledge. (Procedures in scientific method: observation, communication, classification, testing predictions, drawing conclusions, etc.)

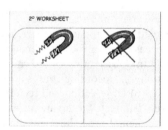

Figure 2. Second worksheet.

3. Developing attraction power with magnets: How many paper clips can a magnet attract?

Materials: (a) Magnets of various shapes, (b) paper clips. Target of activity according to CHAT: (a) Observation of the *child-"subject"* in relation to the magnets and *paper clips* – the "*tools*" chosen to study the power of magnets. (b) Identification and recording of words and phrases which are used by children (language "tools") to describe and explain their observations. (c) Observation of *interactive "subjects"* and "*community*" during the activity. (d) Registration of "*rules*" which are followed during the activity. (e) Observation and recording of "*activity-object.*"

Description of the activity: Children start to observe and experiment with the power of magnets in groups (up to five children). They start by using the magnet and a clip. They bring the *materials* ("*tools*") and make observations. On the first paper clip they place an additional one (forming a chain), exploring in this way the strength of the magnet. The experiment continues, by adding one paper clip at a time to the growing chain, until the magnet no longer attracts others. At the same time the children are given the third worksheet (Figure 3), on which they are asked to represent either a symbol or drawing (symbolic or iconic representation) of the number of paper clips that a magnet attracts.

During the activity, children understand that all magnets have attraction forces to other materials such as paper clips, regardless of their size and shape. In spite of the magnetic force, a magnet will not lose its original properties (attraction, repulsion).

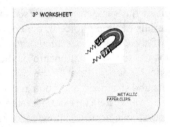

Figure 3. Third worksheet.

4. Developing repulsive forces: the "invisible" force.

Materials: (a) Circular magnets with an empty center, rodlike magnets, horseshoe magnets, compass, wooden base with a rod. (b) Small children's bike. Target of activity according to CHAT: (a) Observation of the children's use of the materials ("tools"-"subjects"-"community"), recording the scientific words and phrases ("tools") which children use. (b) Observation and recording of the methods which are followed by the children-"subjects" to create attractive and repulsive forces. (c) Observation of the experimental procedure and the way they use magnets so as to create repulsive forces ("tools"). (d) Observation of interactive "subjects" and "community" during the activity. (e) Registration of "rules" that are followed during the activity. (f) Identification of the "object" that is constructed during the activity.

Description of the activity: The activity begins by providing children with round magnets and the wooden base with the rod. In addition, we give the children the fourth worksheet (Figure 4), which depicts the final shape that they should create with the magnets and the base. Children begin to carry out the activity in groups (up to five children). More specifically, they begin to lay one to one the circular magnets, using the wooden base. They try to make the magnets stand without touching each other. To carry out this activity children are asked to understand and distinguish the basic concepts of magnetism (attraction, repulsion). In order to build the tower, children must put some given magnets on the base that lies in front of them, with the similar poles of magnets facing each other. The forces thus created must be opposing in order to create repulsion. Then, having experimented with circular magnets, they undertake a similar exercise with other magnets (rodlike magnets, horseshoe magnets, etc.).

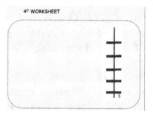

Figure 4. Fourth worksheet.

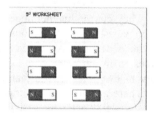

Figure 5. Fifth worksheet.

Figure 6. Sixth worksheet.

In this phase we give children the fifth worksheet (Figure 5), on which they are asked to identify the attraction and repulsive forces that they had just observed. At the end of the activity, the children are encouraged to experiment with magnets and a small child's bike, and to separate the parts of the bicycle which are attracted from those that are not attracted. During the activity, we give children the sixth worksheet (Figure 6) on which we ask them to draw the parts of the bicycle which they observed to be attracted by the magnet.

5. Creating a magnetic field.
Materials: (a) Rodlike magnets, horseshoe magnets, (b) iron filings, (c) paper (size-210x297mm).

Target of activity according CHAT: (a) Observation of the children's use of the materials ("tools"-"subjects"-"community") and recording of the scientific words and phrases (language "tools"). (b) Observation and recording the methods that are followed by the children-"subject" to create the magnetic field. (c) Observation of experimental procedure and recording of the way that children use magnets to create a magnetic field ("tools"). (d) Observation of interactive "subjects" and "community" during the activity. (e) Registration of the "rules" that are followed during the activity. (f) Identification of the "object" that is constructed during the activity.

Description of the activity: In conducting this experiment, we help children understand visually the magnetic field that is created by magnets and that is invisible to them. We provide the children with the necessary materials to create the magnetic field. We encourage them to use magnets freely, in such a way that shows the magnetic force. We ask them to hide the magnet that causes the magnetic field. Using a 210x297mm sized paper, the children then scatter the iron filings in random directions. They put the magnet under the paper, thus creating an invisible magnetic force. Then we ask the children to predict what will happen with the magnetic forces. After the children make predictions, we have a discussion about where magnetic forces grow stronger and explain the reasons why the magnetic forces are stronger at the poles of the magnets. The children verbally express their observations: "There are more iron filings around the poles because there magnetic forces grow stronger." After the activity, we give the children the seventh worksheet (Figure 7), showing three images. Initially, children observe what really happened to the magnet and the iron filings used. The children use the modelling, which is one of the scientific skills and procedures of scientific method. Then, observing the picture of the earth, they are asked to express verbally, using operational definitions, the correlation of magnets with the earth. In the latter figure they note that the center of the earth has accumulated metals, as did the paper on which we put the iron filings.

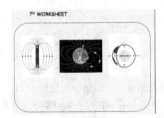

Figure 7. Seventh worksheet.

6. Creating a magnetic compass.
Materials: (a) Rodlike magnets, (b) metal paper clips, (c) round pieces of paper, (d) plastic cup with water.

Target of activity according to CHAT: (a) Observation of how the children use the materials ("tools"-"subjects"-"community") and recording of the scientific words and phrases (language "tools") that children use. (b) Observing and recording the methods the children-"subject" use to create the magnetic compass. (c) Observation of interactive "subjects" and "community" during the activity. (d) Registration of the "rules" that are followed during the activity. (e) Identification of the "object" that is constructed during the activity.

Description of the activity: We provide the children with the materials they will use and the eighth worksheet (Figure 8). The worksheet depicts the procedure that should be followed to create the magnetic compass. We ask the children to rub, in one direction on the magnet, the metal paper clip to create a magnetic field around it. Then the children stick the paper clip on the paper and put the paper on the surface of the water inside the plastic cup. The top of the paper clip will now point north and the back of the paper clip will point the south. We remind the children that magnetic objects such as the metal paper clip are drawn by other metals. We notice that other metallic objects should not be located near the compass that we made. At the end of the activity we ask the children to draw the procedure that they followed to make their own magnetic compass.

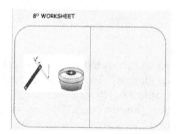

Figure 8. Eighth worksheet.

RESULTS

In this section, we describe a hands-on task in one pre-primary school classroom.

Objectives/Goals: (1) To illustrate the existence of magnetic force and its function. (2) To classify objects and formulate hypotheses regarding materials and their magnetic properties. (3) To illustrate that magnetic force is greatest when the object is closer to the magnet. (4) To illustrate that magnetic force passes through objects; however, the magnetic force decreases with distance. (5) To prove that a magnet can hold a limited amount of weight. During our collaboration with students we stressed the importance of developing, implementing, and studying science teaching and learning based on several principles. Both pupils and university students had multimodal opportunities to engage in different levels of scientific activity, to theorize about the world around us, and to collect and process data (empirical evidence), either in the form of observations or by designing experiments (Roth & Tobin, 2007).

The activity involves two stages: First, pupils in small groups make predictions and classify various materials such as metals, paper, cork, and coins according to how they behave when they approach the magnet. Then, the whole class discusses the categorization of these various materials. Here we focus on the first stage and one particular children's group. Our research question is: How the subject (the university student) used the tools (materials) and how she shared them with different groups of children. The university student used four objects: coins, a piece of paper, plastic objects, and an aluminum can. She asked the children to classify objects according to their magnetic properties. The question was: Which materials can be pulled by a magnet and which cannot? The children divided the tools (materials) into two categories as follows: (1) materials that a magnet can pull and (2) materials that a magnet cannot pull. The study of magnets began with an open exploration of magnetic and nonmagnetic properties of objects (Fig. 1). We provided each pair of students with a horseshoe magnet and encouraged them to go on a "magnetic hunt" with their partner. They explored the room, predicting the behavior of different materials when they came close to a magnet.

The university student asked the pupils to describe a characteristic of each material and the reason why those materials are pulled by a magnet or not. Results are divided in two categories.

1. Video analysis

We videotaped interactions and analyzed them. The total body of data material includes 3½ hours of video material. The goal of our research was to videotape three different groups of students, to evaluate the outcomes with expansive learning and coconfiguration (Engeström, 2005), on one hand, and create new activities on the other. From the activities which the students videotaped, we chose a 10-minute part, based on the eight-step model (Mwanza, 2001). In this stage we note the children's cognitive obstacles.

a. The children do not have a clear opinion about the magnetic materials and the differently shaped magnets.
b. The children cannot notice the difference between the magnetic poles.
c. The children believe that the larger a magnet is, the bigger magnetic force it applies.

Evaluation

Evaluation of videotaped materials in relation to magnetic concepts. At this stage we faced some difficulties, as a significant percentage of students were not familiar with the magnetic properties and with the use of a video camera.

Questionnaires and interviews

The outcomes of the previous stage led us to create a questionnaire and also to conduct an interview with students, so as to learn more about their understanding of magnetic properties. The questionnaire included 13 multiple choice questions.

The first six questions are on content knowledge; the remaining seven questions are about the procedures of scientific methods. The students were classified according to their performance, and then a small group (the top four) were selected to continue to the next stage to ensure that we would not face problems of inadequate content knowledge or cognitive obstacles.

Elaboration of data using the Eight-step Model (Mwanza, 2001)

During work with the four students, they modified the activities and the materials on their own. Furthermore we codefined the schools (community) where students would practice. The four students were divided into pairs and were sent to two different schools for one week. We analyzed a 6-hour video of the students' activities and reached certain conclusions.

Students' school practice

The four students practiced teaching the activities they had already designed. We videorecorded their teaching in order to trace new fruitful activity triangles and evaluate to what extent the outcomes of the previous activities became the objects of new activities and so on.

Reflection

A deep discussion with the full community took place, and we recorded some of the conceptual, social, and emotional impacts of the whole project.

2. Interviews

We interviewed university students/potential early childhood teachers after their 2-weeks of practice in the schools. Students were asked to make comments and find solutions for the planned activities. Two of three pairs explained to us that they faced some problems with the groups of pupils. Many of the children could not focus on the other pupils while they worked with magnets.

They felt more confident with the magnetism concept, and they noticed the opportunities for adopting alternative methods to achieve their aims. Consequently, all pairs reported that pupils faced some problems with the use of the materials. The most important detail was that this theory encouraged different categories of students to participate in groups (pupils from other countries, pupils with hearing problems).

General Assessment of Activities under the Prism of CHAT

To assess the progress of the activities, and the final outcome of the experimental course, it is very important to have the first assessment from the children. We give children the ninth worksheet (Figure 9). The worksheet includes questions that help us to understand if the general experimental course helps children not only to learn about magnetism but also to have a first contact with issues such as magnetism. It is important for teachers-researchers to know if children like the issue of magnetism and also if children develop scientific knowledge of magnetism and about procedures of scientific method.

Figure 9. Ninth worksheet.

Children listen to seven different perspectives and make their decisions based on what they have worked on in the previous days. Then, in response to each sentence they hear, they are asked to draw a smiling face if they agree or an angry face if they disagree. The function of the last worksheet develops metacognitive skills. During the assessment we notice if children have been helped by the experimental course.

At the end of the experimental course we notice that the range of activities helps children develop scientific skills and work collectively and individually as well, according to CHAT. In this course, they have the responsibility to collaborate with children in their group. They are responsible to follow the "rules" of the team-group so as to cooperate better and to establish new "rules" to help their team work more effectively. Children play an active role in the group as "subjects," taking part in the classroom activities. Through the implementation of the activities children develop social skills and take responsibility for their own learning (scientific knowledge), according to CHAT.

They recognize that their goals are achieved through self-confidence, which leads to self-improvement within "community" limits. This experimental course helps to ensure that children:

– Become successful learners who enjoy the procedure of learning.
– Gain confidence about what they can accomplish as "subjects" – through cooperation with the "learning community" and the correct use of "tools."
– Become responsible citizens who make a positive contribution to society/" community."
– Acquire both the "object" of scientific knowledge and the new methods of CHAT.
– Adopt new methods by using CHAT.

The results of the study refer to subject-object orientation and especially focus on object. Furthermore, the various factors of the Engeström's triangle – e.g., rules, community, and division of labor – were studied by analyzing each subtriangle in relation to the major one (see Figure 10).

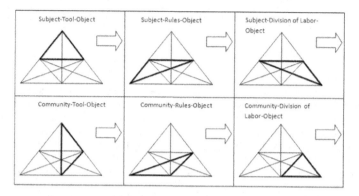

Figure 10. Each subtriangle shows the interaction of three factors that produce different outcomes (inspired by Mwanza & Engeström, 2001, fig. 5 about decompose situation's activity system).

We now proceed with a subtriangle analysis in Table 3. In the left column we describe the activity and in the right column we analyse the activity using a subtriangle CHAT analysis.

Table 3. Situation's activity analysis.

Getting to know about magnets as well as magnetic and nonmagnetic materials	Activity analysis using CHAT
The researcher-teacher helps children use the tools (magnets, magnetic materials, nonmagnetic materials), giving them the opportunity to experiment freely with the materials. Each child chooses the materials which will be used in the activity. Children may work in pairs if they want. In addition they may also cooperate with the researcher/ teacher.	The interaction between the researcher-teacher and the children helps the mediation of the subject-object-tool system, as shown at the triangle below. We provide the children (subjects) with magnetic and non- magnetic materials. We also provide them with different kind of magnets (tools). Children make their predictions for one material each time and write down words using the appropriate terminology (object). The interaction that occurs among children is very important. As the following triangle shows, the interactive system, subject-object-tool, allows both the children and the teacher/researcher to work towards a common goal-object.

175

Subject-Tool-Object

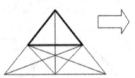

The child experiments freely with the tools (magnets, magnetic materials, nonmagnetic materials). The teacher/researcher suggests that they:

1. Use one magnet each time
2. Choose one material each time
3. Bring together the material with the magnet to observe if magnet can attract or repel it.
4. Put the material that a magnet can attract in a class/group.
5. Put the material that a magnet cannot attract in another class/group.

The interactive system helps to activate and involve children-subjects in the activity and to create rules in the group during the activity. Children (subjects) make rules (use one magnet each time, choose one material each time, put the material that a magnet can attract in a class/group etc) (object).The interaction of the subject-object-rules system is presented in the following triangle. The rules, as well as tools, help achieve the object, as we can see in the previous triangle.

Subject-Rules-Object

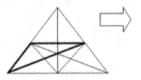

The researcher-teacher helps the group of four children choose the materials and lets them experiment. Then children give instructions to their team members about the way in which they will work (individually or in smaller groups). One child, for example, may be responsible for the management of materials, another for the use of different magnets, another for creating two categories in which materials are placed according to their behavior when contacting the magnet. Finally, a child can keep notes on the worksheet (first worksheet) to record activity progress, and final notes indicating which materials magnets can attract and which they cannot.

In this interactive system (subject-object-division of labor), children-subjects cooperate during the activity and find ways to share responsibilities (division of labor) in each separate group. Some children may be responsible for the materials, some others may use the different kinds of magnets and the rest of the children can keep notes on the worksheet (first worksheet). The children-subjects as a group constantly collaborate with the researcher-teacher, who is involved in the activity only if children request his or her help.

176

Subject-Division of Labor-
Object

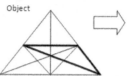

During the activity, the group of children interacts with other groups of children. It is therefore possible to ask some questions of the children in other groups, and also to exchange tools (magnets, magnetic materials, nonmagnetic materials) to determine differences between the magnets used and among different materials.

The triangle depicts the interactive system learning community-tool-object, and it can help the team to interact and share their data and their observations with the other groups. The learning community could be the class or the class and the parents who will participate in a way during the activities. The exchange tools (verbal, materials) help children to collaborate, as they learn to share their materials (magnets, different kind of objects etc), to make different comments, and to exchange their perspectives about the results of each group.

Community-Tool-Object

The groups of children decide which rules will be followed during the activity. They decide to:
1. Use one magnet each time
2. Use one chosen material each time
3. Bring the material close to the magnet to observe if the magnet can attract it or repel it.
4. Put the material that a magnet can attract in a class/group.
5. Put the material that a magnet cannot attract in another class/group.
6. Share and exchange tools and materials with the other groups.

The following interactive system (learning community-object-rules) helps develop the activity. The rules (use one magnet each time, use one chosen material each time, share and exchange tools and materials with other groups etc) that children develop themselves promote better collaboration in each group. Furthermore, the rules ensure the mobility and interaction of the different groups. The different effects-objects in each group provide different aspects of the outcome for the same activity, which leads to new interactive systems.

177

Community-Rules-Object

The groups of children record their observations and the results obtained from the outcome of the activity. They distribute their responsibilities within each group and present the results to other groups. The researcher-teacher encourages the children to work in groups and gives details only if the group requests them. Each group presents the properties of the different magnets that are used and explains to the other groups how they decided to work with the magnets, the magnetic materials, and the nonmagnetic materials.

The interactive system which is presented below (learning community-division of labor-object) develops different functions for the members of each group (division of labor). The division of labor is formed in the wider learning community (community). The learning community is formed by the children participating in the activity and the researcher-teacher (subject), who plays a supportive and cooperative role rather than acting as leader.

Community-Division of Labor-Object

Sort out materials according to their behavior: when approaching the magnet, magnets attract each other and also other objects.	Activity analysis using CHAT
The researcher-teacher helps children use the materials-tools (magnets, glass, wood, metal fasteners, marble, aluminum, copper, brass) by giving them the opportunity to experiment freely with the materials. Each child uses the chosen materials. Children may work together, creating smaller groups. In addition they may also cooperate with the researcher-teacher.	The interaction between the researcher-teacher and children helps, as shown in the following triangle, the mediation of the subject-object-tool. The children (subjects) use the materials (tools) as they wish and try to divide them into two categories. During the activity children observe that traction forces, are a function of the material and are not related to other properties such as shape, size etc (object). The interaction that occurs among children is important and effective, as it allows them to create smaller groups and choose materials and tools they want to use by setting rules. As shown in the following triangle, the interactive

system subject-object-tool allows the children and the teacher-researcher to work towards a common goal-object.

Subject-Tool-Object

The children experiment freely with the materials-tools (magnets, glass, wood, metal fasteners, marble, aluminum, copper, brass). The teacher-researcher proposes:
1. To use one magnet each time.
2. To use one type of material each time.
3. To contact the material with the magnet and to observe whether it is attracted to or repulsed by the magnet.
4. To place attracted materials in a class-group.
5. To place non- attracted materials in another class-group.

The resulting interactive system promotes activation and involvement of the child-subject in the activity and the development of sequence rules within the team during the activity. The interaction concerns subject-object-rules. The children (subjects) use rules (use one magnet each time, use one type of material each time etc) The rules, as well as tools, contribute to gaining the final object, which may be different in each group. The object depends on the interests, desires, and scientific knowledge of the group.

Subject-Rules-Object

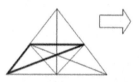

The researcher/teacher helps the group of four children choose the materials with which to experiment. Then he or she lets the children give instructions to their teams about the way in which they will work (individually or in smaller groups). One child, for example, may be responsible for the management of materials, another for the operation and use of different magnets, or another may undertake to create two categories in which materials are placed according to their behaviour when in contact with the magnet. Finally, a child can keep notes on the list that indicates which materials the magnet can or cannot attract. Then the children, as groups or as individuals,

The resulting interactive system (subject-object- division of labor) helps subjects cooperate during the activity and find ways for sharing responsibilities (division of labor) in each individual group. The division of labor in each group depends on the children's ideas about magnetism and about different needs. The children-subjects of each group constantly collaborate with the researcher-teacher, who is involved only if the children request his or her help. The children (subjects) can describe the behaviour of the materials using the

179

may describe the behavior of the materials as indicated on the second worksheet.

appropriate terminology (object). The object could be different for each child and for each group.

Subject-Division of Labor-Object

The groups of children that form during the activity are collaborating. Each team has the opportunity not only to ask questions of the children in other groups, but also to colla-borate with other groups, creating larger groups, sharing tools (magnets, glass, wood, metal fasteners, marble, aluminum, copper, brass).

The interactive system shown in the following triangle (learning community-tools- object) may help the groups interact with and transfer data and observations to other groups. The learning community (children-teacher- class) could transfer and exchange tools (magnets, glass,wood etc). The exchange of tools (verbal, materials) helps children to collaborate, to make different observations, to share the materials they have, and to exchange perspectives after consulting the results of each group. In addition, groups record keywords and phrases pertaining to the subject involved, which helps to obtain scientific know-ledge about magnetism and to use scientific procedures.

Community-Tool-Object

Groups of children set the rules that are followed during the activity. They decide to :
1. Use one magnet each time.
2. Use one material each time.
3. Bring together the material with the magnet and to observe whether it attracted or repelled.
4. Place the tools that are attracted to the magnet in a class-group.
5. Place materials that are not attracted in another class-group.

From the resulting interactive system (learning community-object-rules) it is possible to construct the activity. The rules(use one magnet each time, use one material each time etc) set by children and followed by the groups make cooperation between groups easier and more effective. Furthermore, the rules ensure the mobility and interaction of the different groups that have

6. List on paper the materials in both categories.

emerged. The different effects-objects obtained in each group provide different aspects of the activity's outcome, and then lead to new interactive systems. The learning community within which the activity is taking place is not a separate part of the broader learning community, but an integral part of it.

Community-Rules-Object

The groups of children must write down their observations and the results obtained from the outcome of the activity. Having defined the responsibilities for each group, they present the results to the other groups. The role of the researcher-teacher is supportive, as he or she gives details only at the request of the children. Each group presents the properties of different magnets used and explains to the other groups the way they decided to work with magnets and materials. Moreover the groups present the second worksheet and talk about the properties of the materials using their own epistemology.

The resulting interactive system presented in the following triangle (learning community-division of labor-object) displays the different functions that each group determines for its members (division of labor). Moreover, the division of labor is formed in the wider learning community (community) and in the individual community (groups). The learning community is formed by children participating in the activity and the researcher-teacher (subject), who plays a supportive and cooperative role rather than leading. The learning community can then be revised, and more children or children's classes can be added.

Community-Division of
Labor-Object

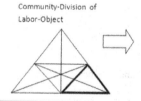

Developing Attraction force	Activity analysis using CHAT
The researcher-teacher helps children process the materials (metallic paper clips, magnet), giving them the opportunity to experiment freely with them. Each child individually chooses the material that will be used in the activity. Children may also cooperate with	The interaction between the researcher-teacher and the children helps to improve functioning of the subject- object-tool system, as shown below. The children (subjects) start to observe and experiment with the

other children and with the teacher-researcher.

power of magnets (tools). Children start with a magnet and a paper clip, they bring the materials and make observations. During the activity children understand that all magnets have traction forces to other materials such as paper clips (object). The interactions that occur among children are also important.

Subject-Tool-Object

While children are experimenting freely with the materials (metallic paper clips, magnets), the researcher-teacher suggests that they:
1. Use one magnet each time.
2. Use one metallic paper clip, which connects to the previous.
3. Bring together the magnet with the paper clips.
4. Stop adding clips when the magnet cannot pick up others.

The resulting interactive system (subject-object-rules) facilitates the activation and involvement of the children in the activity, and in some cases the team sets rules during the activity. Specifically children working on groups try to set rules (use one magnet each time, use one metallic clip, which connect to the previous, etc). At the end of the activity children understand that in spite of the magnetic force, a magnet will not lose its original properties (traction, repulsion) (object).

Subject-Rules-Object

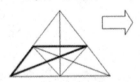

The researcher-teacher helps the group of four children put forward the materials. The researcher-teacher encourages the children to give instructions to their group regarding how they will work. One child is responsible for the paper clips, another for the operation and the use of the magnet, the third child counts paper clips placed on a chain which the magnet pulls, and the fourth child keeps notes on the worksheet (third worksheet) to record

In the resulting interactive system (subject-division of labor-object) children work together during the activity and find ways to share responsibilities in each individual group. Children in groups constantly collaborate with the researcher-teacher, who will participate only if his or her help is requested by the children.

the progress of the activity and the final number of paper clips that the magnet can lift.

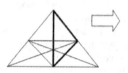

Subject-Division of Labor-Object

During the activity the group of children interacts with other groups of children. Therefore it is possible not only to ask questions of children in other groups, but also to share tools (magnets, paper clips) and to find the differences between the magnets used initially.

The resulting interactive system (learning community- tool- object) helps each group interact and transfer data and observations to other groups. The exchange of tools helps children work together, make different observations, share material, and exchange perspectives on the results-outcomes of each group.

Community-Tool-Object

Groups of children record the rules that must be followed during the activity. They decide:
1. To use one magnet each time.
2. To use a paper clip, which will then connect to another clip.
3. To touch the magnet with the paper clips.
4. To stop adding paper clips on the chain when the magnet cannot pick up others.
5. To share and exchange tools and materials that have been used by other groups.

The activity is constructed from the resulting interactive system (learning community-object-rules). The rules set by the children make cooperation in each group easier and more effective. The rules ensure the mobility and interaction of the different groups that have emerged. The different effects–objects obtained in each group provide different aspects of the outcome of the same activity, which then lead to new interactive systems.

Community-Rules-Object

The groups of children must record their observations and the activity outcomes. They distribute the responsibilities within each group and present the results-outcomes that they have obtained to other groups. The help of the researcher-teacher is supportive, as he or she gives details only at the request of the groups. Each group of children presents their

In the interactive system learning community-division of labor–object, different responsibilities are observed. The division of labor is observed in the wider learning community. The learning community is formed by the children participating in the activity, and the researcher-teacher plays a

findings-observations about the strength of magnets used and explains their findings to the other groups that the method they decided to work with was magnets and paper clips.

supportive and cooperative role rather than a leadership role.

Community-Division of Labor-Object

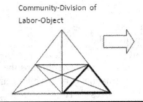

Development of repulsive forces – the invisible force	Activity analysis using CHAT

The researcher-teacher helps children use the materials (circular magnets with an empty center, rodlike magnets, horseshoe magnets, compass, wooden base with a rod), giving them an opportunity to experiment freely with the tool- materials. Children may work together, creating smaller groups, and with the researcher-teacher.

The interaction between the researcher-teacher and children provides the mediation of the subject-object-tool system, as shown in the following triangle. The interaction between children (subjects) is important, as it allows them to create smaller groups, choose materials and tools, and to edit or set rules. As shown in the following triangle, the interactive system subject- object-tool allows
the children and the teacher-researcher to work towards a common goal-object.

Subject-Tool-Object

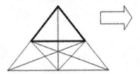

The children experiment with the materials (circular magnets with an empty center, rodlike magnets, horseshoe magnets, compass, wooden base with a rod). The teacher-researcher proposes that they:
1. Use a pair of magnets each time.
2. Observe the behavior of magnets when the magnets contact each other (attraction, repulsion).
3. Put the circular magnets on the wooden base so they do not come into contact with each other (repulsion).

The interactive system facilitates the activation and involvement of the children-subject in the activity. The interaction concerns subject-object-rules system. The rules, as well as tools, help achieve the final object, which in each group could be differ-ent, depending on the interests, desires, and scientific knowledge of the children. The rules set by each group form both the outcome of the activity and the final result-object.

4. Create a "tower" (wooden base with circular magnets) in which "rocks" (circular magnets) are not in contact.

Subject-Rules-Object

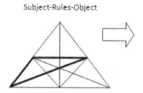

The researcher-teacher helps the group of four children experiment with materials. Then he or she encourages the children to instruct their teams about the manner in which they will work (individually or in smaller groups). One child, for example, may be responsible for managing magnets, another for the proper placement of the magnets, while another child plans to depict the behavior of each side of the magnets when they are in contact. Children as a group or as individuals may plan and place different letters on the sides of the magnets (fourth worksheet). Moreover, groups of children may experiment with the two poles of differently shaped magnets and make observations. Children confirm their observations on the fifth worksheet. They observe the attraction and repulsive forces between the magnets regardless of their shape. In conclusion we let children experiment with a small child's bicycle and magnets of different shapes. Children paint the parts of a bicycle which are observed to attract in the sixth worksheet.

In the resulting interactive system (subject-division of labor-object), children-subjects cooperate during the activity and find ways to share responsibilities (division of labor) in each individual group. The division of labor in each group depends on the children's ideas, not only in relation to magnetism, but also in relation to the different needs of groups. The children- subjects constantly collaborate with the researcher-teacher, who is involved only if his or her help is requested by the children. The final object could be different, not only for each child, but also for each group. The division of labor in each group enhances the cooperation between the subjects.

Subject-Division of Labor-
Object

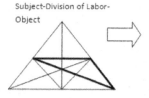

During the activity, groups of children collaborate. Each group has the opportunity not only to ask questions of the children in the other groups, but also to collaborate with other groups, thus creating larger groups, sharing tools, and making observations.

The interactive system, learning community-tool- object, enhances the interaction of the group and the transfer of data and observations between groups during the activity. The exchange of tools (verbal, physical) encourages children to collaborate, to make different observations, to share materials, and to exchange perspectives on the results of each group. In addition, groups record keywords and phrases for the object of the activity and develop scientific procedures.

Community-Tool-Object

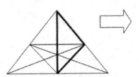

The groups of children record the rules that must be followed during the activity. Each group will decide to:

1. Use a pair of magnets each time.
2. Observe the behavior of magnets when in contact with various materials (attract, repel).
3. Put the circular magnets on the wooden base so they do not come into contact with each other (repulsion).
4. Create a "tower" (wooden base) in which "rocks" (circular magnets) are not in contact.
5. Choose and place a different symbol for both poles of magnets.
6. Observe the parts of the bicycle that magnets attract.
7. Paint the parts of the bicycle that magnets attract.

The resulting interactive system, learning community-rules-object, shapes the activity. The rules created by the children make cooperation in each group more effective. Furthermore, the rules ensure the mobility and interaction of different groups, delimite the way in which each group works. The different outcomes-object provide different aspects of the outcome of the activity and lead to new systems of interaction. The learning community within which the activity takes place is an integral part of the broader learning community.

Community-Rules-Object

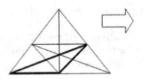

The groups of children write down their observations and results obtained from the outcome of the activity. Having defined the responsibilities of each group, they present the results- outcomes to the other groups. The role of the researcher-teacher is supportive, as he or she gives details only at the request of the groups. Each group presents the properties of different magnets used and explains to the other groups the way they decided to work with magnets and materials.

The following interactive system (learning community-division of labor- object) displays the different functions that each group assigns to its members (division of labor). The division of labor formed in the wider learning community (community) consists of the children participating in the activity and the researcher-teacher (subject), who plays a supportive and cooperative role, rather than a leadership role. The learning community may be revised and may involve more children or children's groups.

Community-Division of
Labor-Object

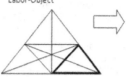

Creating a magnetic field	Activity analysis using CHAT
The researcher-teacher helps children use the tools (magnets, iron filings), providing an opportunity to experiment freely with them. Each child chooses the material and works alone or in collaboration with other children, creating smaller groups. The collaboration with the researcher-teacher is important in this activity, as researchers observe difficulties of pupils in management concepts (magnetic field) and in the management of materials.	The interaction between the researcher-teacher and children improves the functioning of the subject-object-tool system. The interaction that occurs among children is important because it allows them to create smaller groups to develop materials and tools or to set rules. As shown in the following triangle, the interactive system subject-object-tool allows both children and the teacher-researcher to work towards a common goal-object.

Subject-Tool-Object

The children experiment with the tools (magnets, iron filings). The teacher-researcher proposes that they:

1. Use the magnet and the filings without being in direct contact with them.
2. Create a visible magnetic field with the iron filings, the paper (A4 size), and a magnet.
3. Observe the lines of the magnetic field that were generated.
4. Observe where magnetic forces and magnetic lines are stronger.

The interactive system promotes the activation and involvement of children-subjects in the activity and the development of the sequence rules within the group during the activity. The interaction ensures that rules, as well as tools, contribute greatly to the group's ability to obtain the final object and to make observations about the magnetic field created.

Subject-Rules-Object

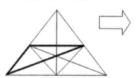

The researcher-teacher helps the group of four children use the materials-tools. Then he or she encourages the children to give instructions to their groups about how to use the material (individually or in smaller groups). One child, for example, may be responsible for the management of materials, another for the use of magnets, another may be responsible for dropping the iron filings on the paper. A child holding the magnet underneath the paper (which is not to be seen) develops the magnetic field. Finally, a child takes notes on the magnetic field that is created.

The interactive system subject-object-division of labor helps children-subjects cooperate during the activity and find ways for sharing responsibilities (division of labor) in each individual group. The division of labor in each group depends on the children's ideas about magnetism. The children-subjects of the group constantly collaborate with the researcher-teacher, who is involved only if the children request his or her help. The researcher-teacher's contribution during the activity is important, as the children are trying to develop concepts that are not obvious to the children.

Subject-Division of Labor-Object

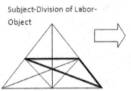

The groups of children taking part in the activity collaborate. Each group has the opportunity not only to ask questions of the children in other groups, but also to collaborate with other groups, creating larger groups, sharing tools (magnet iron filings) and mainly exchanging perspectives on the creation of the magnetic field.

The interactive system shown in the learning community-tool-object triangle could help the group interact and transfer data and observations from other groups. The exchange of tools (verbal, materials) helps children collaborate and, most importantly, make observations and exchange views about the results. In addition, groups note keywords and phrases relevant to magnetism. This helps both groups gain scientific knowledge and learn scientific procedures.

Community-Tool-Object

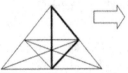

Groups of children record the rules that must be followed during the activity. They decide to:
1. Use the magnet and the filings without being in direct contact with them.

The resulting interactive system, learning community-rules-object, helps children set rules for the groups to follow. The rules make cooperation

2. With the iron filings, the paper (A4 size-210x 297mm), and a magnet, create a visible magnetic field.
3. Observe the magnetic field lines that are generated.
   Observe where magnetic forces and magnetic lines are stronger.
4. Note the similarities and differences between the magnetic lines observed in iron filings and the image of the magnetic lines surrounding the Earth (seventh worksheet).
5. Observe where stronger magnetic forces occur.

Groups of children record their observations and results based on the activity outcome. The children define the responsibilities of each member of the group. The groups present their results to the other groups. The role of the researcher-teacher is supportive, as he or she gives details only at the request of the children. Children in the group present the properties of the created magnetic field and the magnetic field of the earth.

in each group easier and more effective. Furthermore, the rules ensure the mobility and interaction of different groups. The learning community within which the activity takes place is not a separate part of the broader learning community, but an integral part of it.

Community-Rules-Object

The interactive system learning community-division of labor-object presents different functions for its members (division of labor), which are determined by each group. The division of labor formed in the wider learning community (community) and the individual (groups) is formed by the children participating in the activity and the researcher-teacher (subject), who plays a supportive and cooperative role, rather than a leader-ship role. The learning community may then be revised and more children or children's groups may be added to it.

Community-Division of Labor-Object

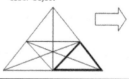

Magnetic compass-orientation	Activity analysis using CHAT
The researcher-teacher helps children use the tools (magnets, paper clips, piece of paper, plastic cup with water) and gives them the opportunity to experiment freely with the materials. Each child uses the materials and chooses to work alone or in collaboration with other children, creating smaller groups. Working with the researcher-teacher is important in this activity, as children face	The interaction between the researcher-teacher and children helps facilitate the interaction of the subject-object-tool triangle, as shown below. The interaction that occurs among children is important because it allows them to create smaller groups in which to develop materials-tools and to set rules. As shown in the

189

difficulties in the management of materials.

following triangle, the interactive system subject-object-tool allows children and the teacher-researcher to work towards a common goal-object.

Subject-Tool-Object

The children experiment with the materials-tools (magnets, paper clips, piece of paper, plastic cup with water). The teacher-researcher encourages them to:

1. Use the magnet and the paper clip to construct a magnetic needle.
2. Place the magnetic needle in the water to build a magnetic compass.
3. Observe the magnetic forces around the magnetic compass.

The interactive system facilitates the activation and involvement of the children-subjects in the activity and the development of sequence rules during the activity. The interactions of the subject-object-rules system ensure that rules, as well as tools, contribute to achievement of the final object.

Subject-Rules-Object

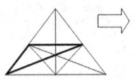

The researcher-teacher helps the group of four children use the materials. Then he or she encourages the children to instruct their groups about how to use the materials (individually, in smaller groups). Finally a child keeps notes on the magnetic compass and how the group uses the materials.

The resulting interactive system, subject-object- division of labor, encourages children- subjects to cooperate during the activity and find ways to share responsibilities (division of labor) in each individual group. The division of labor depends on the children's ideas about mag-netism and magnetic compasses. The children-subjects of the group const-antly collaborate with the researcher-teacher, who is involved only when asked to help. His or her contribution during the activity is important.

Subject-Division of Labor-Object

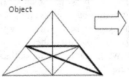

The groups of children that form during the activity collaborate. Each group has the opportunity to ask questions of the children in other groups and to collaborate with other groups, thus creating larger groups, sharing tools (magnet connectors), but mainly to exchange perspectives on the creation of the magnetic compass.

The interactive system shown in the learning community-tool-object triangle shown below could help the group interact with other groups and transfer data and observations. The exchange of tools (verbal, materials) helps children to collaborate, make observations, and exchange perspectives on the results of each group. In addition, groups record keywords and phrases are exposed to scientific procedures such as modeling.

Community-Tool-Object

The groups of children define the rules that are followed during the activity. They decide to:
4. Use the magnet and the paper clip to construct a magnetic needle.
2. Place the magnetic needle in the water to build a magnetic compass.
3. Observe the magnetic forces deployed around the magnetic compass.

From the resulting interactive system (learning community-object-rules), it is possible to construct the activity. The children set the rules that are to be followed by the group, which makes cooperation easier for each group. Furthermore, the rules ensure the mobility and interaction of the different groups that have emerged. The learning community of the activity is not a separate part of the broader learning community, but an integral part of it.

Community-Rules-Object

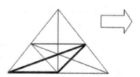

The resulting interactive system, which is presented in the triangle of learning community-division of labor-object, displays different functions for each group (division of labor). In addition, the division of labor is formed in the wider learning community (community) and the individual (groups) ensures the cooperation of

191

the children involved in the activity and of the researcher-teacher (subject), who plays a supportive and cooperative role, rather than a leadership role. The learning community can then be revised, and more children or children's groups can be added to it.

Community-Division of
Labor-Object

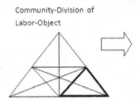

Conclusions and Implications

In conclusion, the results of using Cultural-Historical Activity Theory (CHAT) for science education laboratory lessons on magnetism seems promising. Cultural studies of science education seem to have the potential to make science education available to all citizens or, in other words, an alternative way to scientific literacy. The theoretical framework of expansive learning (Engeström, 2005) seems to be appropriate and fruitful for science education researchers. The Mwanza eight-step model seems to be more easily understood by science education learners. The students' capability to design learning activities could offer criteria for analyzing and evaluating learning in university science education laboratories.

In an expansive cycle, decisive actions of individuals within an activity system, which emerges in reaction to and resolution of deep internal contradictions, coalesce to form a qualitatively new mode of joint activity. An expansive cycle can be thought of as the reorchestration of an activity system that occurs when there are shifts within and between its six "corners" (Engeström, 1999).

In recent decades there has been considerable progress in developing a European model of society-"community" that is unique in the world. It is based on both economic growth and providing good jobs for people-"subjects." Although this model ensures a high level of social protection and education, the educational system remains static in many cases. CHAT is a method which could provide both "community" and education as well. CHAT uses elements of traditional education such as books (tools) as a means for education. Moreover the traditional classroom becomes a dynamic field of learning both in the traditional place of education (school) and in different learning environments such as museums, libraries, and even outdoor areas such as parks. This different way of using traditional "tools" and methods could develop and stabilize CHAT in the educational system.

We note that Europeans have grown up with the benefits of this social model, which consisted of "rules," "division of labor," and mediation between "subjects" and "object." It is therefore very important for children to learn new methods for working more efficiently.

Studying the results of PISA 2006 (Programme for International Student Assessment) highlights the need for a fruitful new framework to achieve the goal of scientific literacy. In our view, this need may be fulfilled by CHAT. It is important for the Greek educational system and for other European educational systems to develop new educational strategies.

With this experimental course we seek to enhance education and especially science education and training, starting at younger ages. Achieving these results is very important for researchers of science education who conduct research on new methods such as CHAT and who incorporate new "tools" to achieve the "object," as we realized from the activities analysis using CHAT. More than anything else, CHAT provides a new beginning, a social pillar to modernize the European social model by investing in human capital and social inclusion and by moving towards science education research that focuses on structural change (Jahreie & Ottesen, 2010).

This case study, even though limited, could push forward the boundaries of teacher training in science education. The project was effective, and the outcomes resulted from the contribution of several groups. First of all, school teachers who teach natural sciences in the early grades saw this didactical intervention as a chance to improve science teaching in their classes. Also, collaboration with the students at the workshop provided an opportunity to investigate the way the approach works in a real classroom situation and to discuss their data collection with other students, as well as with nursery school teachers. The most important result was that university students were then able to design activities following the eight steps of the Mwanza model 2001).

NOTES

[1] This research has been co-financed by the European Union (European Social Fund – ESF) and Greek national funds through the Operational Program "Education and Lifelong Learning" of the National Strategic Reference Framework (NSRF) – Research Funding Program: Heracleitus II. Investing in knowledge society through the European Social Fund.

REFERENCES

Brown, A. L. (1992). Design experiments: Theoretical and methodological challenges in creating complex interventions in classroom settings. *Journal of the Learning Sciences, 2*(2), 141–178.

Cole, M. (1995). Socio-cultural-historical psychology: Some general remarks and a proposal for a new kind of cultural-genetic methodology. In J. V. Wertsch, P. D. Río, & A. Alvarez (Eds.), *Socio-cultural studies of mind* (pp. 187–214). New York: Cambridge University Press.

Cole, M. and the Distributed Literacy Consortium. (2006). *The fifth dimension: An after-school program built on diversity.* New York: Russell Sage Foundation.

Engeström, Y. (1987). *Learning by expanding: An activity-theoretical approach to developmental research*. Helsinki: Orienta-Konsultit.

Engeström, Y. (1999). Activity theory and individual and social transformation. In Y. Engeström, R. Miettinen, & R.-L. Punamäki (Eds.), *Perspectives on activity theory* (pp. 19–38). New York: Cambridge University Press.

Engeström, Y. (2005). *Developmental work research: Expanding activity theory in practice*. Berlin: Lehmanns Media.

Hasu, M., & Engeström, Y. (2000). Measurement in action: An activity-theoretical perspective on producer-user interaction. *International Journal of Human-Computer Studies, 53*(1), 61–89.

Hedegaard, M., & Fleer, M. (2008). *Studying children. A cultural-historical approach*. London: Open University Press.

Jahreie, C. F., & Ottesen, E. (2010). Construction of boundaries in teacher education: Analyzing student teachers' accounts. *Mind, Culture, and Activity, 17*(3), 212–234.

Kaptelinin, V., & Nardi, B. (2006). *Acting with technology: Activity theory and interaction design*. Cambridge: MIT Press.

Kaptelinin, V. (1996). Activity theory: *Implications for human-computer interaction*. In B. A. Nardi (Ed.), *Context and consciousness: Activity theory and human-computer interaction* (pp. 103–116). Cambridge, MA: The MIT Press.

Kuutti, K. (1996). *Activity theory as a potential framework for human-computer interaction research*. In B. A. Nardi (Ed.), *Context and consciousness: Activity theory and human-computer*. Cambridge: MIT Press.

Mwanza, D., & Engestrom, Y. (2003). *Pedagogical adeptness in the design of elearning environments: Experiences from Lab@Future project*. Paper presented at the E-Learn 2003 International Conference on E-Learning in Corporate, Government, Healthcare, & Higher Education, Phoenix, AR, 7–11 November.

Mwanza, D. (2001). Where theory meets practice: A case for an activity theory based methodology to guide computer system design. In M. Hirose (Ed.), *Proceedings of INTERACT'2001: Eighth IFIP TC 13 International Conference on Human-Computer Interaction*, Tokyo, Japan, July 9–13. Oxford: IOS Press, UK INTERACT 2001. Download paper as KMi Technical Reports version in PDF KMI-TR-104.

Nardi, B. (Ed.). (1996). *Context and consciousness: Activity theory and human-computer interaction*. Cambridge: MIT Press.

OECD. PISA 2009 results. *What students know and can do. Students performance in reading. Mathematics and science*, Volume I.

Plakitsi, K. (2008). *Science education in early childhood: Trends and perspectives*. Athens: Patakis [in Greek].

Plakitsi, K. (in press). *Sociocognitive and sociocultural approches in early childhood*. Preface by Wolff-Michael Roth. Athens: Patakis [in English and in Greek].

Roth, W.-M. (2005). *Doing qualitative research*. Rotterdam: Sense Publishers.

Roth, W. M., & Lee, S. (2004). Science education as/for participation in the community. *Science Education, 88*, 263–291.

Roth, W. M., & Tobin, K. (2007). Science, learning, identity. In *Sociocultural and cultural-historical perpectives* (pp. 203–204). Rotterdam: Sense Publishers.

Roth, W.-M., Hwang, S., Lee, Y-J., & Goulart, M. (2005). *Participation, learning, and identity: Dialectical perspectives*. Berlin: LehmannsMedia.

Theodoraki, X., & Plakitsi, K. (2009). Activity theory and learning in science education laboratory lessons. The case of magnetism. In G. Cakmakci & M. F. Tasar (Eds.), *Contemporary science education research: Learning and assessment* (pp. 207–213). Retrieved from http://www.esera2009.org/books/ Book_4.pdf.

van Eijck, M., & Roth, W.-M. (2007). Keeping the local local: Recalibrating the status of science and Traditional Ecological Knowledge (TEK) in education. *Science Education, 91*, 926–947.

Yin, R. K. (1994). *Case study research: Design and methods* (2nd ed.). Applied Social Research Methods Series. Thousand Oaks, CA: Sage Publications.

Xarikleia Theodoraki
University of Ioannina Greece
Greece

Katarina Plakitsi
University of Ioannina
Greece

ELENI KOLOKOURI & KATERINA PLAKITSI

8. A CULTURAL HISTORICAL SCENE OF NATURAL SCIENCES FOR EARLY LEARNERS

A Chat Scene

INTRODUCTION

This paper is part of a wider study concerning scientific literacy and understanding of the Nature of Science (NOS) from the early grades (AAAS, 1989; NRC, 1996; UNESCO, 2008). It is connected with activity theory in formal and informal science education, which aims to contribute to an emergent agenda about cultural-historical activity theory (CHAT) and science education in Europe. This project leads to rethinking scientific literacy (Roth and Lee, 2004), in the sense that it develops new methodological tools that could reform science education. Scientific literacy has become a priority for educational experts and institutions in many European countries and worldwide. UNESCO having declared 2003–2012 as United Nations Literacy Decade is developing an "Information for All" program, in which science awareness is considered vital (UNESCO, 2008). Moreover, increasing literacy by 2015 has been connected with poverty reduction according to the Summit on the Millennium Development Goals (UNESCO, 2010). An appropriate science teaching curriculum at all educational levels promotes understanding of natural concepts and develops scientific argumentation concerning various natural phenomena. Thus, future citizens will lead a life of responsibility and decision making in a contemporary society. An important tool, which many scientists propose for teaching science in the early grades, is studying Nature of Science. NOS offers a humanistic approach to science, as it helps beginners realize that science is a human activity with social applications (Bravo, 2005). Moreover, prospective teachers become familiar with teaching methods that will help them overcome their anxiety about teaching science and design stimulating classroom activities. An innovative method of introducing natural concepts in the early grades is the use of interdisciplinary means such as comics and cartoons-animations (Bongco, 2000). A series of research studies worldwide has shown that using animations in classroom activities encourages creative thinking, stimulates children's interest, combines knowledge with everyday actions, and finally enhances understanding of concepts in different fields, as well as concepts of natural sciences (Keogh and Naylor, 1999; Yang, 2003). The aims of the study concern both prospective teachers and early-grade pupils as they refer to assessment of the effects of an alternative teaching method on prospective teachers, understanding of natural concepts in the early grades, creation of science learning

K. Plakitsi (ed.), Activity Theory in Formal and Informal Science Education, 197–228.

environments connected with CHAT, and contributions to scientific literacy. Following the adventures of SpongeBob SquarePants, prospective teachers approach the concepts of floating and sinking in a workshop designed for this purpose. They make predictions about which objects float or sink, test their predictions, provide reasons, and try to overcome their cognitive obstacles. Moreover, they practice the skills of scientific method and finally they design classroom activities according to NOS basic principles. As a next step, they teach floating and sinking concepts in preprimary school classrooms using techniques of drama in education in combination with science education techniques. The research data collection follows a variety of methods used in a case study and seeks to prove the appeal that this teaching strategy has in pre-primary school classrooms. Results confirm that this innovative practice (1) enhances understanding of natural concepts in the early grades, (2) improves science teaching, (3) enables both university students and pupils to deal with basic scientific concepts and express themselves using their own epistemology, (4) directs the professional development of prospective teachers, (5) supports collateral learning, (6) makes NOS a horizontal axis in both natural and social sciences (7) stresses the importance of cultural-historical activity theory (CHAT) in science education (8) situates science education as a part of society, and (9) pushes forward the boundaries of scientific literacy. It is our belief that scientific literacy leads to sustainable development and environmental awareness, which are vital in the contemporary world. Thus, we invest in the sociocultural background of the citizen who will learn science, develop a positive attitude towards nature and the environment, and also make decisions about scientific and technological matters. Furthermore, by using basic principles of CHAT in classroom situations we contribute to the establishment of science education as participation in the community, as well as·a new ontology and epistemology in teaching scientific concepts.

THEORETICAL BACKGROUND-FRAMEWORK

Scientific literacy has become a priority for educational experts and institutions in many European countries and worldwide. From a sociocultural perspective (Vygotsky, 1978; Roth and Lee, 2004), it is important to provide learners of all educational levels with an appropriate science teaching curriculum in order to achieve scientific literacy. Understanding natural concepts and developing scientific argumentation concerning various natural phenomena will lead future citizens to a life of responsibility and decision making in contemporary society. According to Miller (1983), "in a democratic society, the level of scientific literacy in a population has important implications for science policy decisions" (p. 29). His efforts concentrated on defining scientific literacy, contributing to its measurement in the United States and Europe, examining the results in 34 countries, and discussing the implications of the results for education policy, science policy, and democratic government in the 21st century (Miller, 2007). Laugksch (2000) refers to four groups of scientists and researchers who promote scientific literacy as a whole or as a part of the community. These "interest groups" include the science education community, social scientists and public opinion researchers concerned with scientific and technological issues, sociologists of science, and science educators concerned with a sociological approach to scientific literacy. UNESCO has declared 2003–2012 as United Nations Literacy Decade and is developing an "Information for All" program, in which science awareness is considered vital (UNESCO, 2008). Within the framework of this action they stress the importance of designing curricula that provide a multidisciplinary approach to science and technological education and basic knowledge and scientific literacy for all. Accordingly, AAAS and NRC have taken the initiative to advance scientific literacy and modern aspects of Nature of Science (AAAS, 1989; NRC, 1996). Moreover, increasing literacy by 2015 has been connected with poverty reduction according to the Summit on the Millennium Development Goals (UNESCO, 2010). A 10-year (2006–2015) framework entitled "The Literacy Initiative for Empowerment (LIFE)" has been developed, in which effort is put into supporting literacy in 35 countries of high challenge (UNESCO, 2007). From all the above we reach the conclusion that scientific literacy is connected with the needs of all citizens, and this makes it a collective activity. In this sense, scientists and citizens interact in a dynamic system in order to gain knowledge and achieve scientific literacy within the sociocultural framework. Furthermore, scientific literacy is a concept that can occur "in the wild-that-is-in the everyday world that we share with others as opposed to testing situations in classrooms and laboratories" (Eijck & Roth, 2010, p. 1). This study uses Cultural Historical Activity Theory (CHAT) as a theoretical framework in order to design and analyze natural science education activities towards scientific literacy. In this regard, teaching natural sciences becomes a dynamic activity system which involves multiple participants (university teachers, lab assistants, students, early grade pupils, etc.). All the participants act towards some common goals, considering scientific knowledge as a cultural, historical, and social process and using mediative and analyzing tools.

According to Bravo (2005), it emerges as a social imperative that all people need to know science and also know about science. Moreover, scientifically literate citizens who are aware of selected topics from the philosophy of science make decisions about science and technology. An important tool that many scientists consider vital for science education in the early grades is studying Nature of Science (NOS). Scientists share certain beliefs and adopt a series of attitudes connected with nature of science. Many researchers (Abd-el-Khalick, Bell, & Lederman, 1997; Hogan, 2000; McComas, 1998, 2005) have studied various aspects of NOS and its application in natural sciences curricula and scientific literacy projects, as well as teachers' perspectives on NOS. Furthermore, the development of certain NOS principles (Lederman, 1992; McComas, 1998, 2005) creates a new field in the design of scientific curricula. These principles are a powerful methodological tool for designing natural science activities in the early grades. NOS application offers a humanistic approach to science, as it helps beginners realize that science is a human activity with social applications (science as culture). Taking this into account, this study is based on using NOS to promote scientific literacy from the early grades. At an international level great importance is placed on the child's education, even from the first steps of his or her life. The fact that infants try to explore their surrounding environment by using their senses and try to satisfy their natural curiosity provides a foundation for developing scientific concepts. As international research documents point out (AAAS, 1989; NAEYC, 2002; NRC, 1996; UNESCO, 2008) education in natural sciences must start at a very early stage. At this level, learning in natural sciences is connected with exploring in authentic environments, practicing skills of observation, classification, communication, etc., and making sense of the world around us. Reform of the Greek national curriculum for the early grades in 2003 has put science education on a cultural-historical foundation. The content of science education in the early grades concerns topics and concepts from the surrounding environment: human life, social structures and relations, the life of plants and animals, places, and people, and natural phenomena. One of the central goals is to make pupils understand the relations and interactions that exist in the world we live in by engaging them in proper activities. Within this framework, pupils are considered as a member of a family, of a neighborhood, of a community in the village or city they live in (p. 4, Greek National Curriculum for the early grades).They lead their lives in a socially organized community with all the offers, demands, and conflicts they meet under the influence and protection of their family. According to Hedegaard (2008), neither society nor its institutions (i.e., families, kindergarten, school) are static; rather, they change over time in a dynamic interaction between a person's activities, institutional practice, and societal traditions. When the pupil enters the school institution (classroom) for the first time, he comes face to face with another socially organized environment in which he has to adapt through participating in everyday activities. The role of the teacher is that of a mediator who has to take into consideration all the different cultures in a classroom and thus facilitate the learning procedure in a pedagogical, interactive, and discursive atmosphere.

Science educators agree that not everybody needs to know how to repair a car, how to build a house, or how to grow plants, but most of them insist that everybody needs to obtain basic knowledge of scientific concepts (Roth & Lee, 2004). These concepts will be a guide to leading a life of responsibility and decision making in a technologically advanced world. Consequently, prospective teachers feel more confident to teach natural sciences and make decisions concerning scientific matters (Lederman, 1992; McComas, 1998; Plakitsi, 2007).

An innovative method of introducing natural concepts in the early grades is using interdisciplinary means such as comics and cartoons (Bongco, 2000). Educators have stressed the importance of using art in classroom activities, as taken from the work of Dewey (1934), who supports every child's right to art, to Gardner (1983), who promotes different types of intelligence (e.g., spatial intelligence in the visual arts). Horn (1980) states that "cartoons have the ability to make a point without the semantic ambiguities inherent in the written words." He also characterizes cartoons as a universal language that develops the imagination of the readers. Keogh and Naylor (1999) support teaching and learning scientific concepts by using concept cartoons. Thus, they offer an alternative method of science teaching which involves discussion, investigation, and motivation for learners of all educational levels. Yang Gene (2003), in his website Comics in Education (http://www.geneyang.com/comicsedu/), presents the strengths of comics in education as motivating, visual, permanent, intermediary and popular. Scott McCloud (1993) sees a different perspective: "In learning to read comics we all learned to perceive time *spatially*, for in the world of comics, *time and space are one and the same*" (p. 100). Humor, exaggeration, symbols, and emotions are all elements that provide learners with very interesting types of knowledge presented in a familiar context. It is a fact that both comics and cartoons-animations are related to ideas we already know or have directly experienced, which makes learning concepts meaningful (Mayer, 1996). Furthermore, they engage early starters in exploring a variety of scientific concepts, in experimenting, in creative thinking and providing solutions to problems without restraint. For example, the cartoon program "Dora the Explorer" seems to have an impact on pupils concerning geographic education, as it offers a variety of skills and capabilities that can be a basis for learning geography, orientation, and developing problem-solving strategies (Carter, 2008). Research about using comics and cartoons as an educational tool is widespread. The British Cartoon Archive (in British Cartoon Archive http://www.cartoons.ac.uk/teachingaids) encourages research about cartoons and supports groups of people who wish to exchange views about using cartoons in learning and teaching and also get ideas and inspiration. Accordingly, in the web page of Comic Life (in Comic Life http://plasq.com), visitors can discuss comics in education and find information to create their own stories.

The New York City Comic Book Museum (http://www.nyccomicbookmuseum. org/education.htm) offers educators an opportunity to bring comic books into the classroom. They become involved in comics literacy, the history of comics, and useful information for creating comic books. At this point we would like to stress the importance of the sociocultural aspect of teaching natural sciences in the early grades and address the following research questions:

1. Does CHAT in the context of natural sciences education help university students become capable of teaching natural concepts in the early grades?
2. Does it provide motivation to prospective teachers to develop innovative natural sciences activities for their pupils?
3. Is CHAT a theoretical framework suitable for meaningful learning and scientific literacy development in the early grades?
4. Is it possible to reform natural sciences education by using CHAT?

PURPOSE

The aims of this research study were set under the perspective of considering scientific knowledge as a dynamic activity system in which the participants, the institutions, the methods, the tools, the objects are connected in a cultural, historical, and social process. Moreover, scientific knowledge overcomes its individual identity and becomes a collective activity with applications in science and society. In this sense, the study seeks to:

1. Use Cultural-Historical Activity Theory (CHAT) as a theoretical framework in order to design and analyze natural sciences education.
2. Assess the effects of an alternative teaching method on prospective teachers.
3. Determine the extent to which a certain teaching method enhances understanding of natural concepts in the early grades.
4. Provide motivation to teachers and prospective teachers for developing innovative science activities for their pupils.
5. Provide motivation to teachers and prospective teachers for developing creative and critical thinking as well as collaborative skills for their pupils.
6. Provide motivation to prospective teachers and pupils for becoming acquainted with research skills.
7. Provide early starters with the foundation for meaningful learning and literacy development.
8. Provide early starters the opportunity to practice skills of the scientific method.
9. Develop positive attitudes and values towards natural sciences.
10. Contribute to scientific literacy and decision making of prospective teachers.

RATIONALE

The rationale for this study is based on the observation that both teachers and prospective teachers appear to be not very well informed about teaching natural sciences in the early grades. This makes them hesitant to adopt suitable didactic strategies for introducing the content knowledge and applying didactic transformations. According to Roth and Tobin (2002), both prospective and in-service teachers found university teacher training was not preparing them adequately for their actual classroom teaching, indicating that there is a need to bridge the gap between theory and practice within the framework of teacher education. The proposal addressed in this study is the introduction of the sociocultural aspects of teaching natural sciences in the early grades as a means of reforming natural sciences education. In other words, using CHAT in the context of natural sciences education in order to motivate prospective teachers to develop innovative natural sciences activities for their pupils and help them teach natural concepts in the early grades. Thus, CHAT becomes a theoretical framework suitable for meaningful learning and scientific literacy development.

Considerable research (e.g., Bravo 2005; Lederman, 1992; Plakitsi, 2007) on teacher education has reported the fact that teachers have their own ideas about science and the Nature of Science. These ideas can be different from the scientific views and sometimes seem to cause anxiety to teachers. As a result, they need to become familiar with teaching methods that will help them overcome worrying images of science and design stimulating classroom activities. Of the NOS principles, we have stressed the importance of the following five:

2. Knowledge production in science shares factors, habits of mind, norms, logical thinking, and methods.
5. Science has a creative component.
6. Science has a subjective element.

7. There are historical, cultural, and social influences on the practice and direction of science.

9. Science and its methods cannot answer all questions.

These principles provide a powerful methodological tool for designing natural sciences activities in the early grades. This study presents an application of NOS in the classroom in which certain abilities are practiced that are useful for exploring the concepts of floating and sinking. The production of scientific knowledge is a result of various factors. Initially, by using their senses, prior knowledge, and cause-effect reasoning, students are led to make predictions about the behavior of materials when they are in water. They then make hypotheses, test them through experimenting, and finally make their decisions about floating and sinking concepts. In this way, scientific knowledge, which includes subjective elements such as personal beliefs, is tested. For example, during the workshop practice, even after experimenting, students could hardly believe that an orange floats, as they had difficulty releasing the cognitive obstacle that the weight of an object does not define its floating and sinking behavior.

METHODOLOGY

The methodology plan used both in working and interacting with the university students and in the application in preschool classrooms is based on:

(1) The framework of analysis by Yrjö Engeström (2005).
(2) The cultural-historical approach of Marilyn Fleer and Marianne Hedegaard (Hedegaard & Fleer, 2008, Fleer & Hedegaard, 2010) about children's development in everyday practices.
(3) The project "The Fifth Dimension," implemented by Michael Cole and the Distributed Literacy Consortium (2006).

Engeström's theoretical framework of analysis seems to be appropriate and fruitful for natural sciences education researchers. The cultural-historical approach of Marilyn Fleer and Marianne Hedegaard is a useful guide for university students, as it provides all the information about children's development in a sociocultural environment and the skills needed to design innovative natural sciences activities. Finally, Michael Cole and the Distributed Literacy Consortium have provided, in "The 5th Dimension Project," an example of an educational activity system in which university students can see multiple interactions of subjects, objectives, tools, rules, division of labor, etc., taking place in different educational settings. All those interactions of different methodological processes could offer criteria to analyze and evaluate learning in university natural sciences education. In this regard, the learning process focuses on connecting cognitive structures – not with the individual but with culture. Thus, the analysis of learning activities demonstrates the way scientific concepts and phenomena are incorporated into the sociocultural framework of the learner. Instead of providing model activities which must be followed, we provide tools which the learner can use to mediate the learning process. Taking this into account, we try to invest in the potential of the learner, who can reach new cognitive structures while experimenting. Consequently, a new mentality is developed in which the new methodological processes become a strong tool that teachers can use to push the boundaries of scientific thinking and learning in the classroom.

The study is divided into three parts:

First Part

The first part concerns a workshop within the frame of a university lesson entitled "Didactics of Natural Sciences," which includes a series of natural science activities and the proper teaching strategies to teach floating and sinking concepts. Prospective teachers attended the workshop and practiced the following sections:

1. Narration of the adventures of SpongeBob SquarePants (in http://www.sponge bob-games.com/), a popular cartoon figure who lives in a city under the sea and who faces unexpected problems of floating and sinking.
2. Predictions about which items float or sink.
3. Testing of predictions.
4. Providing reasonable causes for floating or sinking.
5. Overcoming cognitive obstacles (or pre-conceived notions).
6. Identifying skills of the scientific method.
7. Designing classroom activities.
8. Connecting teaching strategies with NOS.

Second Part

The second part concerns the didactic scenario which was designed following the basic principles of CHAT (group work, use of instrumental and conceptual tools, interactions between subjects, mediation between subjects and community). The didactic strategies used included certain techniques of drama in education in combination with science education techniques. Prospective teachers had an active role in designing the activities in the workshops. They thus became able to adopt suitable teaching strategies in order to teach floating and sinking concepts in pre-primary classrooms. As prospective teachers moved from one stage to the other, they exchanged roles in order to find the solution to problems concerning floating and sinking concepts. They defined place and time, and through role-playing, argumentation, the conduct of experiments, and evaluation they reached the conclusion. As a result of this collective activity, prospective teachers became engaged in designing natural sciences activities in a sociocultural environment and studied the interactions in the preprimary classrooms of Ioannina.

The letter. Prospective teachers receive an envelope which contains a letter from SpongeBob. It is accompanied by photos of Bikini Bottom and of Bob and his friends, a sticker album, and a magazine in which students can see the adventures of SpongeBob. The letter describes everyday life in Bikini Bottom, which was normal until the day the wicked witch Lavinia the Maze changed the rhyme of

SpongeBob SquarePants' song, as well as the substance of water. As a result, certain parts of the city started to float while others sank, and Bikini Bottom was led to destruction. It is in Bob's hands to save the city, so he asks for help. Looking in his great-grandmother's books he finds a plan. He has to find out which items float and which ones sink in water and build a model city using these materials.

Teacher in role. The teacher in the SpongeBob role discusses the problem with prospective teachers and provides information about the city and the situation described in the letter. Prospective teachers ask questions and try to find a way to help Bob save his city. Possible questions from prospective teachers to teacher/SpongeBob: How did you come here? How do you and your friends feel about the witch? Are you the same Bob that we watch on TV? What will happen to the city now? In what way can we help you?

Painting. Prospective teachers draw the city of Bikini Bottom on a big piece of paper. This is the fictional place that will help them play roles and feel more comfortable in the drama situation.

Argumentation. Prospective teachers are divided in two groups: the floating group and the sinking group. Each group has to discuss the behavior of certain materials when put in water and present argumentation as to why some of them sink and others float. A representative from each group announces estimations and provides reasons. They make predictions about the behavior of each material in water and then they fill in the following board of predictions.

Prediction board. Material: stone, nail, button, potato, orange … Sinks: YES … NO … Floats: YES … NO …

Table 1. Prediction board

Material	Sinks	Floats
Stone		
Nail		
Buttons		
Cork		
Sponge		
Tomato		
Pepper		
Pineapple		
Cucumber		
Eggplant		
Potato		

Prospective teachers use their senses to describe the materials and predict what will happen if they put them in the water.

Experiment. Prospective teachers put the different materials one by one in the water and observe what happens. They classify the materials in two categories according to their behavior in the water. Finally they test their predictions and discuss the cognitive obstacles and the skills of scientific method used, and they provide ideas for extra activities. Moreover, they try to find out what ideas and cognitive obstacles preschool pupils have about floating and sinking and discuss them in groups.

Telephone conversation. Prospective teachers listen to one part of a telephone conversation. The teacher in the role of Bob receives a telephone call from Patrick, his best friend in Bikini Bottom. Bob writes down four key phrases that he hears from Patrick and tries to find out, with the aid of prospective teachers, what they mean. Each phrase leads to an experiment which is performed in class.

Phrase 1: A whole peanut or half of it floats or sinks–you will see!

Phrase 2: Put peanuts in carbon dioxide, and they will perform amazing tricks by your side!

Phrase 3: Cut the potato, cut the potato!

Phrase 4: Step in a boat of plasticine and travel away through the ocean.

Talking about the phrases one by one, prospective teachers conduct the experiments in groups.

They make a list of the materials needed, predict the behavior of the materials, organize and conduct the experiments. At the end of each experiment they test their predictions and discuss the cognitive obstacles, the skills of scientific method used, and the ideas and cognitive obstacles preschool pupils have about the situations in each experiment.

Evaluation. On a sheet of paper divided into two horizontal parts, prospective teachers draw the items that float on top and those that sink to the bottom. These drawings contain useful information about the behavior of the materials when put in the water, so they will be sent to SpongeBob.

Game. Prospective teachers find a way to make their racing boats move in water without touching them.

Frozen Pictures. Prospective teachers present scenes of Bikini Bottom city using their bodies. The teacher in role asks every group to talk about the scene they have presented.

Discussion in circle. The teacher who played the role discusses the knowledge they have obtained so far with prospective teachers, as well as the prospects of saving the city of Bikini Bottom. Prospective teachers in groups discuss the activities, the utility of a popular cartoon character in natural science activities, and the application of the didactic scenario in a pre-primary classroom.

Third Part

The third part of our study is an application of the previous activities in preprimary classrooms. The research plan for prospective teachers is to follow the steps they have already practiced during the workshop and teach floating and sinking concepts in seven classrooms with approximately 25 pupils each. The research data collection follows the methods of a case study, which are extensively applied in social sciences and in research education (Cohen & Manion, 1994). Two external observers record the evolution of the didactic strategies, the emergence of spontaneous ideas from pupils, and the classroom interactions. A videotape recording of the classroom activities provides material for discussion and evaluation of the teaching process. Finally, a semistructured interview with prospective teachers responsible for each classroom is another source of evidence in which, in addition to the questions answered, suggestions are made and solutions are proposed.

Seven pairs of third-grade prospective teachers were identified from a sample of 110, all of whom participated in the workshops concerning a variety of scientific concepts. Selection was based on questionnaires designed to explore prospective teachers' scientific knowledge about floating and sinking concepts as well as mini-projects and discussions about socioscientific issues. Prospective teachers had the opportunity to teach floating and sinking concepts over the course of two weeks in seven different pre-primary school classrooms. (Several pupils were highly interested in learning about other adventures of SpongeBob and his friends, so we decided to continue the study for about two more weeks.) Prospective teachers had little previous experience in real classroom situations, so they were willing to put the knowledge and skills they had obtained into practice.

The pre-primary schools in which the study was conducted were situated in the area of Ioannina, and each classroom had 20 to 25 pupils. Two children with hearing problems were reported, three with difficulty in speaking, and four of foreign origin, all of whom participated actively during the whole procedure.

Teachers in the schools collaborated with prospective teachers in providing all the relevant information about their science program and the children's cultural and social background. They also provided them with materials needed for the activities and encouraged them to prepare teaching activities on a low budget. Nursery school teachers were generally enthusiastic about exchanging views on scientific matters with both students and our research team.

Five members of our research team played the role of the observer in order to record the evolution of the didactic strategies and the emergence of spontaneous ideas from pupils. The classroom activities were videotaped, photographs were taken, and we all met at the end of each day in order to evaluate the process and prepare for the following session.

Videotape recordings of the classroom activities provided material for discussion and evaluation of the teaching process and the interactions in it. Finally, a semistructured interview with the prospective teachers responsible for each classroom was conducted in which, in addition to the questions answered, suggestions were made and solutions were proposed.

The didactic scenario for all pre-primary school classrooms was designed in connection with the Cross-Thematic Curriculum Framework for Nursery School. Emphasis is put on "the cross-thematic perception, the holistic perception of knowledge and the development of the interests and the ideas expressed by children while learning" (Cross-Thematic Curriculum Framework for Nursery School, 2003, p. 253).

The development of communicative skills, collaborative and creative work, problem solving, and critical thinking are some of the priorities of the curriculum's planning and development. The planning and development of the context of different subjects such as language, mathematics, studies of the environment, drama, music, and physical education are not considered as independent actions; all the subjects interact during implementation in the classroom. Moreover, the Indicative Funda-mental Cross-thematic Concepts in the curriculum support pupils' learning with skills, attitudes, and values which reinforce cross-thematic perceptions. Taking this into account, teaching scientific concepts is an interdisciplinary procedure that takes place not only in the school classroom but also in the laboratory and in the environment with strong connections to society. Learning is an ongoing process which is affected by societal conditions, while aims and goals can be modified according to current circumstances and pupils' interest.

Table 2. Connection with the Greek National Curriculum (5–6 years old. Cross-Thematic Curriculum Framework for Nursery School, 2003, p. 253.

Content Guiding Principles	General Goals (knowledge, skills attitudes and values)	*Indicative Fundamental Cross-thematic Concepts*
With appropriate teaching interventions, both at the nursery school and the wider environment, pupils should be provided with many opportunities to discover the basic characteristics of the properties of different materials.	Pupils are encouraged to observe and find the similarities and differences between the materials they use. They are also encouraged to classify materials according to common properties such as color, transparency, behavior in water, etc. (studies of the environment, mathematics, music, art).	Individual-Group Interaction System Classification

RESULTS

Results Referred to Preparatory Training of Prospective Teachers

Questionnaires

The questionnaires contain two parts: In part A (see Appendix B) students express their scientific views in an extract of the research instrument VOSTS (Views on Science-Technology-Society) (Aikenhead, 1989). The questions concern the definition of science and the scientific method, science in the community, understanding of science, social lives of scientists, the role of errors in the advance of science, etc. In part B they provide the right answers about sinking and floating concepts related to skills of the scientific method. The pairs that were finally selected provided views which showed that they had attended most of the STS, that is Science-Technology-Society lessons and workshops, during their studies and gave the right answers to part B of the questionnaire. Moreover, these students seemed to have considered learning not just an individual action but a matter of collaborative activity in which the interactions of different elements and people played an important role.

At this point, it is interesting to comment on the most common answers that students provided in the first part of the questionnaire. It seems that emphasis is put on answers that reveal the social aspect of science and scientists and on the connection and collaboration between science and society. Thus, we can see that science is defined, in most cases, as a discovery of new things in the world in which we as a community live; technology has a great impact on the way we think and interacts with science in order to facilitate progress. The connection between science and the arts reveals another strong bond between science and society. Questions, hypotheses, data collection, and conclusions are tools collectively used to advance science, while errors are believed to be a source of learning that can lead to new discoveries. Theories are tested many times and for long periods by the community of scientists before they are established as laws. As far as scientists' personal lives are concerned, it seems that upbringing, intelligence, ability, a natural interest in science, as well social and family status are all elements that comprise a scientist's personality. In other words, a scientist is not just a talented individual isolated in his or her laboratory but a member of a community; his or her progress is a result of multiple interactions that take place inside the community. Finally, the way a country trains its scientists seems to have an impact on the way they look at problems and on the solutions they provide.

Interviews

The interviews were conducted with the pairs of prospective teachers that implemented this didactic scenario in class. Questions at this point (see Appendix B) concerned the experience of teaching and playing different roles, the attitude of pupils towards the cartoon and towards the whole didactic scenario, evaluation of the use of cartoons in a classroom and connecting them with natural sciences, etc. Answering the questions one by one, prospective teachers revealed a more confident orientation to handling scientific views in class. They faced some difficulty in role playing and insisted on having more practice with it. Using cartoons in science activities made them feel that they had a powerful means for keeping pupils' interest alive during the whole procedure. They also felt familiar with alternative teaching methods that will help them design stimulating classroom activities. Finally, they felt more confident about investing in their future profession as they gradually became capable of overcoming their anxiety and dealing successfully with scientific matters in class. The re-examination of previous workshops through the prism of CHAT provided insights that may have been missed otherwise. The theory helped them to formulate interesting research questions in order to design activities, to set goals not necessarily connected to the pupils' previous knowledge but to the interactions and conflicts that take place in a school classroom. Thus, implementation of the didactic scenario directed attention to the use of tools, the division of labor, the role of the teacher, and the interactions within the community. In this way a much deeper understanding of the context of natural sciences was established that involved the implementation of new didactic strategies and evaluation processes which have brought forward innovative aspects of science. Within this theoretical framework, the prospective teachers internalized the basic principles of CHAT and, almost without prompting, acquired a new mentality about teaching natural science concepts.

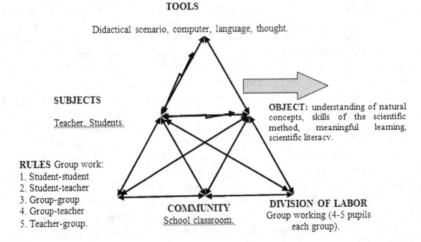

Figure 1. The mediational structure of an activity system (Engeström, 1999) Sample model of the implementation of the didactic scenario within the CHAT framework.

The tools used in this activity system (didactic scenario, computer, language, thought) mediate with the subjects (teacher, students), so that students become able to reach the object (understanding of natural concepts of floating and sinking, skills of the scientific method, meaningful learning, scientific literacy). The learning community (school-classroom), sets rules for the members of the community and the groups (pupil-pupil, pupil-prospective teacher, group-group, group-prospective teacher, prospective teacher-group), forms flexible groups of four or five pupils, and divides the labor. The object/ outcome of the activity develops positive values and attitudes towards natural sciences, encourages the use of skills of the scientific method, and finally leads to scientific literacy within the modern sociocultural frame. In this regard, the learning procedure is based on the multiple interactions that take place in the community and supports the development of tools such as drama, cartoons, and role playing, all of which help pupils learn how to combine their personal experience and knowledge with the interactions of the group. Thus, understanding of natural concepts is a result of collaborative action, critical thinking, problem solving, and argumentation through the prism of modern sociocultural theories.

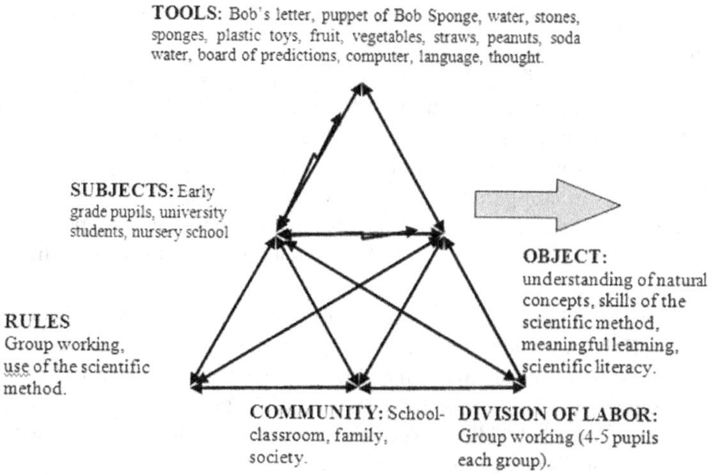

TOOLS: Bob's letter, puppet of Bob Sponge, water, stones, sponges, plastic toys, fruit, vegetables, straws, peanuts, soda water, board of predictions, computer, language, thought.

SUBJECTS: Early grade pupils, university students, nursery school

OBJECT: understanding of natural concepts, skills of the scientific method, meaningful learning, scientific literacy.

RULES Group working, use of the scientific method.

COMMUNITY: School-classroom, family, society.

DIVISION OF LABOR: Group working (4-5 pupils each group).

Figure 2. The mediational structure of an activity system (Engeström, 1999). Sample model of the students' implementation of CHAT in the classroom.

In this activity system, the multiple mediations that take place (tools, rules of interaction, division of labor) are responsible for the balance of the activities in it. Tools mediate with subjects and help them reach the object. The learning community follows rules by which groups of four or five pupils are formed, and thus the division of labor becomes more flexible. Finally, the object includes learning outcomes such as understanding of natural concepts of floating and sinking, meaningful learning, and scientific literacy. Contradictions or change of interactions between mediations affect the whole activity system. For example, when a prospective teacher is confident while playing the role of SpongeBob, dealing with classroom disturbances is much easier as he can find ways to modify the use of tools or the division of labor in the activity. This is effective in keeping the balance inside the activity system and, as a result, he succeeds in working collectively towards the common goal and reaching the CHAT perspective's definition of the activity, which is not "doing" but "doing in order to transform something" (Engeström, 1987).

Results Referred to Early Grade Pupils

The results of the video analysis follow the steps of the didactic scenario used to teach floating and sinking concepts in the seven preprimary classrooms.

The letter.   In all seven classes the arrival of a letter from a popular cartoon hero was met with mixed feelings of enthusiasm and wonder about the contents of the envelope. Pupils were already familiar with the hero, as they had been watching his adventures on television. Most of them sang the song along with the teacher and provided specific information about the city of Bikini Bottom and citizens other than SpongeBob. They found it hard to believe that SpongeBob was actually addressing their class, but when they realized that he corresponded with other children as well through his magazine, this disbelief was overcome. The appearance of a witch had a great impact on them, and apart from the suggestions they made for helping Bob they offered to make her extinct! Finally they provided some initial information about the floating and sinking behavior of several items in Bikini Bottom. One pupil confirmed that sponges are living things, as she had seen plenty of them on her father's island last summer and offered to bring some photos to the classroom so as to compare them with SpongeBob. From all the above, we can see that pupils connect scientific concepts with their own experience and knowledge, which makes learning a collaborative activity with social applications.

Teacher in role.   Initially there was a lot of hesitation about the teacher in the role of SpongeBob, but when the teacher explained that this was part of a game of exchanging roles in order to help SpongeBob, they were willing to participate. This explanation appealed to two of the classes, while in other classes pupils wanted to play the role of SpongeBob. The questions/suggestions that they made to the teacher in role are of great significance, as they reveal not only the pupils' knowledge about the properties of water but also provide information about their language skills, ability to express opinions, decision making, geography, pollution of the environment, etc. We provide the following examples of pupils' questions: Where is the Pacific Ocean? – Is the water very cold down there? – How are you able to breathe and talk on the sea bottom? – What language do you speak? – Why is your nose so different from ours? –Why does the jellyfish stick to you? – Do you know that certain people take sponges like you from the sea bottom and sell them in shops? – Is the sea bottom polluted where you live? – How did you travel to Greece? – What do you want us to do? – We are willing to help you save your city! Finally, in one class, the pupils stated clearly that they already know materials that float and others that sink but that experimenting would help a lot.

Painting.   Painting the city of Bikini Bottom was the result of group work, role playing, and decision making in all classes as the pupils slowly created the place where the hero and his friends live. This activity gave pupils the chance to improve their creativity and imagination, as well as oral expression. The pupils with hearing problems, as well as those with difficulty in speaking, participated equally and were encouraged to do so by the whole class.

Argumentation-prediction board. In this part pupils were encouraged to express their opinions about the behavior of each material when put in water and to fill in the prediction board. They did so by using pictures which represented each material, since at this stage most of them have not fully conquered reading and writing skills. They also used alternative vocabulary to refer to floating and sinking, such as "it swims," "it goes up," "it falls down the water," which were accepted because at such an early age children must be encouraged to describe scientific concepts any way they can.

Experiment. The experiment was the crucial point in all classes, as it provided the information needed for helping SpongeBob and was conducted with great interest. Pupils saw their predictions prove true or false and provided explanations for each case. Moreover, they tried to come close to the scientific truth and became little scientists themselves.

Pupils' scientific explanations.

The potato is heavy so it sinks.

The candle does not sink because it is light.

The tomato is in the water, but it does not sink, it swims.

The wooden toy floats because it is made of wood and wood floats.

If we put a lifejacket around the potato it must keep it on the surface because it has air inside.

At this point the pupils' observations show their efforts to provide scientific explanations and open the field for further experiments.

Telephone conversation. This drama technique created a lot of enthusiasm and expectation in all classes, as pupils could hardly believe that a cartoon character had actually called them to give further instructions. The imaginary situation led them to act, to try to make sense of the four phrases, and to deal with the scientific contents collectively. While reading each phrase aloud, they decided on further action following the same plan each time: they brought the appropriate materials, made predictions, conducted the experiments, and reached conclusions about the behavior of each material when it was placed in the water. The experiments that were conducted at this point showed that pupils had obtained a basic vocabulary of floating and sinking concepts (e.g., it sinks, it floats, the air inside makes it swim, the potato has a dense body), as well as an ability to organize the experiment by suggesting what they should do, bringing the appropriate materials, and providing explanations.

Evaluation – discussion in circle. At this final stage there was a retrospection of the whole procedure, which showed that the scientific terms used by pupils to describe floating and sinking concepts indicated they had obtained basic knowledge and that they had developed skills for creative thinking and problem solving. They talked about floating and sinking concepts in connection with all the materials that can be found in SpongeBob's city and drew instruction leaflets for Bob. Finally, they talked about the difficulties they faced, as well as their misconceptions about the behavior of the materials. For example, most of the pupils thought that big materials sink; however, they found out this was not true and that they had to consider factors such as the shape, density, and hollow parts of the material.

Table 3. CHAT Interpretation of several episodes of the practical application.

Transcription of the science lesson	*Interpretation*
– A pupil finds the letter of SpongeBob on the doorstep of the school. She delivers it to the university student wondering who could have sent it and how.	Pupils connect the fact that they have received a letter with an ordinary activity. Their parents receive letters at home so the letter becomes a tool to motivate pupils to act as a group dealing with a situation.
– The university student reads the letter in class. Pupils hesitate to believe that a cartoon character has sent them a letter.	Pupils lead their life in a socially organized community under the influence and protection of their family. Learning that the cartoon character has his own personal life makes them believe that learning is a collective activity, as they have to deal with scientific concepts in order to collaborate with the cartoon.
– Pupils confirm that they have seen the cartoon character on TV, on clothes, and toys and ask for details about his residence, his friends, his birthday, and his daily life. They wonder if TV characters lead daily lives as they do.	Chrysanthi confirms that sponges are living things, as she has seen plenty of them on her father's island last summer and offers to bring some photos to the classroom so as to make a comparison with SpongeBob.
– Pupils draw Bikini Bottom, the place where the hero and his friends live in groups.	In pupils' drawings, represent-tations of natural phenomena are obvious: some drew the city on the sea surface because they wanted it to float, others said that an earthquake caused severe damage and drew everything upside down, while others presented drawings of an ordinary home like their own.
– A university student asks if pupils know the meaning of floating and	Pupils talk about scientific concepts in connection with personal experience

sinking. They answer that something floats when it is on the sea and it can breathe when it does not go down. Something sinks when it goes down in the water, and it cannot come out.

– Pupils provide examples of floating materials (ship, bucket, the island of Ioannina, lifejacket, armbands, shells, clothes, swimming suit) and sinking materials (stones, sand, keys, cell phone, watermelon, the Titanic).

as well as previous knowledge. Learning is taken out of the classroom, as science exists in the community (school, home, holidays, films, etc.) in everyday life.

– A university student provides pupils with a variety of materials and encourages them to explore their properties by using their senses.

Pupils explore the materials on their own and with other pupils and ask the university student for more information.

Group interactions among university student, pupils, and materials (tools) take place during the activity.

– Pupils stand in two rows, the sinking and the floating row. Those from the first row pick up items that sink and provide a reason for this behavior, while those from the second do the same with items that float. If someone cannot decide, he/she stands between the two rows and listens to the arguments of both rows so as to decide about the behavior of the item.

Pupils use argumentation in order to talk about the behavior of different items when put in water. They provide reasons for floating and sinking based on the experience of their own sociocultural background, they use their own epistemology in order to describe the natural phenomena, and finally they try to persuade those who have not decided yet about floating and sinking behavior. Pupils interact as a group so as to decide about the behavior of the items. University student provides help only if asked.

– In order to help SpongeBob save his city they make suggestions: we could use only items that sink so they can stay at the sea bottom; we could tie the items that go on the surface with rope; we could put Mr. Plankton in prison and hire another witch to make things as they were before Lavinia came.

Pupils participate in a problem-solving situation and interact with other pupils and university students in order to reach the desired outcome, which is to help their cartoon friend.

– While experimenting with the behavior of an egg in salty water, pupils suggest that they should visit

Pupils connect the behavior of an egg in salty water with swimming in the sea in contrast to swimming in a pool.

the sea and do the experiment there.	They refer to a force that pushes you up in the sea which makes swimming easier than in a pool (buoyancy).
– Pupils provide reasons for sinking or floating according to the weight, size, or texture of the material.	Pupils describe scientific concepts, providing examples of their logical thinking and their everyday life.
– After the experiment Dimitra wanted to make the potato float. She put it inside a hollow plastic toy and said: "See? I have made a lifejacket for the potato, so it can swim!"	Pupils describe scientific concepts and modify them in a creative way.
– Pupils listen to one part of a telephone conversation. They hear four key phrases and try to find out what they mean. Each phrase leads to an experiment which is performed in class.	Pupils organize the experiments in class, they interact within the group and with the university student, and they set rules and use new tools in order to conduct the experiments.
– When it was time to say goodbye to Bob, pupils asked Bob how he would go back to the Bikini Bottom. They wanted to find the city on the map; they found the Pacific Ocean with the aid of the university students.	Pupils show interest in extending the knowledge they have gained and suggest the use of new tools (map, computer).
– Pupils asked nursery school teachers to visit Bob's site on the Internet and play games.	

This interpretation was inspired by Marianne Hedegaard and Marilyn Fleer's analysis (2008) of children's development in everyday practices. This study presents a modification of their analysis, which combines the child's perspective with the epistemology of natural sciences in early childhood.

CONCLUSIONS – IMPLICATIONS

Implementation of this study reveals that scientific literacy is a collective property involving lifelong participation and learning (Roth & Lee, 2004). In this sense, several groups contributed to the results of this study. School teachers who teach natural sciences in the early grades saw this didactic intervention as a chance for improving science teaching in their classes. They suggested that their ideas about science teaching in combination with the curriculum and the connections of science to society would help them overcome worrying images of sciences and would contribute to their professional development. Collaboration with the students in the workshop has shown that adopting interdisciplinary means such as comics and cartoons-animations bridges the gap between science education and public knowledge and awareness about science.

Comics and cartoons as a form of art influence and reinforce the learning process by using symbols and pictures. Humor and exaggeration, familiar characters, and objects from daily life seem to have appealed to university students and pupils. Dealing with scientific concepts with the aid of a popular cartoon character contributed to better understanding of science, helped to make connections with prior knowledge, and helped build a strong interactive network in order to achieve meaningful learning of the scientific content. Furthermore, university students had the opportunity to investigate the way it works in a real classroom situation and discuss their data collection with other students, as well as with nursery school teachers. As a result, preservice and in-service teachers collaborated with researchers in designing and analyzing natural sciences activities, used the same conceptual tools, collected important data about their own work and the interactions with others, and finally found ways to deal with conflicts in the activity system. Conducting research through the prism of CHAT led us to consider the pupil's perspective in relation to the activity, as well as the sociocultural situation in each school. The motives, personal interests, and aims of pupils interacted dynamically with all the people they worked with in a specific institution.

Following the CHAT theoretical framework, collaboration within the activity system has been proved successful, as the framework provided a variety of opinions regarding the use and effectiveness of mediating tools. Problem solving was a result of group work in which both verbal and nonverbal communication were used in order to reach the solution. Division of labor resulted from collective action and responsibility in the sense that researchers, workshop assistants, university students, nursery school teachers, and pupils exchanged knowledge and experience while working together. Special needs were accepted and dealt with in the community. Aims were set under the perspective of considering scientific knowledge as a dynamic activity system and were adapted in several cases according to the demands of the community. Thus, goal setting and process of learning were interpreted as open collective activities in a specific sociocultural frame. Participation at all levels was major, and learning outcomes were connected with pupils' interests and sociocultural backgrounds.

It seems that at the end of the whole procedure the three dimensions that Miller recognizes in scientific literacy were reached: (1) an understanding of the norms of the methods of science (i.e., the nature of science), (2) an understanding of key scientific terms and concepts (science content knowledge), and (3) an awareness and understanding of the impact of science and technology on society (Miller, 1983). These three dimensions have been reinforced with pupils and university students working on the scientific method. Practicing skills of the scientific method, such as observation, classification, prediction, experiment, communication, and interprettation, became a useful tool for moving towards scientific literacy. Data analyses of the questionnaires and discussions about the mini-projects have shown that scientific literacy is achieved by taking an active role in learning. Furthermore, we are quite certain that this method is much more motivating than the traditional ones, as it is closer to everyday life. In-service and prospective teachers play an important role in this process, as they encourage children to engage in interactive learning. As Hedegaard (2008) stresses while describing the role of teacher: "It's not like, you will learn, this is what I'm teaching you, but doing something together, looking something up together, adding some information, that's a really important part of the process" (p. 79). It is important to consider not only what a teacher delivers to a pupil in the process of teaching but also what teacher and pupil do together that brings in new ideas and new ways of thinking. In this regard, the teacher takes an intermediary role that allows him or her better understanding of the pupil dealing with scientific concepts. This makes learning natural concepts meaningful and at the same time keeps away worrying images which hinder progress.

On the grounds that interdisciplinary and innovative teaching is a priority for education in the early grades, this study contributes to engagement with science and pushes forward the boundaries of scientific literacy. At the end of this study, we feel certain that we have provided pupils with a strong knowledge base that will become the foundation for meaningful learning and scientific literacy achievement. It is our belief that scientific literacy leads to sustainable development and environmental awareness, which are vital in the contemporary world. Thus, we invest in the sociocultural background of the citizen that will learn science, develop a positive attitude towards nature and the environment, and also make decisions about scientific and technological matters. It is our duty to build on this from an early age and to intervene in students' thoughts so as to help them approach scientific truth.

Furthermore, we provide teachers and researchers with new orientations, such as using tools in order to create innovative activities and studying the interactions in an activity system that relates everyday concepts to scientific content. By using basic principles of CHAT in classroom situations we contribute to the establishment of science education as participation in the community, as well as a new ontology and epistemology in teaching scientific concepts. This innovative practice makes NOS a horizontal axis to both natural and social sciences as it directly affects teachers' training. Prospective teachers had an intermediary role in the school class-room towards making connections with society, culture, and the environment. As a result, knowledge production in science has proved to be a creative component of methods, thinking, interactions, and social practices. NOS implies cultural and social influences on the practice and direction of science, and these influences have been evident in this study, as knowledge of the scientific content has been combined with interpretation of experiences and information throughout daily life. In conclusion, we would like to stress the importance of the fact that prospective teachers were offered the opportunity to reflect on their teaching of natural concepts in the classroom. Thus, they were able to see how pupils construct their knowledge, what epistemological views of science they present while doing science, and the connections they make between science and daily life. This in-depth understanding of what prospective teachers are expected to teach helps them to support the development of pupils' understanding of NOS. Implications and recommendations for further research concerning this study suggest a holistic understanding of the world by means of education, science, culture, and communication. Furthermore, the research could indirectly contribute to interconnections between action research methods and sociocultural ones.

APPENDIX A

The Letter

Dear friends,
In the deep blue sea, very far away from here, in the Pacific Ocean, there is the sea city of Bikini Bottom. Although it is built underwater, it is almost like every other city with streets, houses, schools, cinemas, playgrounds, an airport, etc. The sea bottom resembles the earth's surface in this city, so the citizens of Bikini Bottom lead a normal life, are able to breathe, move around, entertain themselves, etc., just exactly like earth citizens.

This is the city where I live in a house in the shape of a pineapple, and my name is SpongeBob SquarePants. I have a lot of friends and neighbors down there: Squid, Sandy, Patrick, and others. Every morning Pat the pirate sings a rhyme and the day starts.

The rhyme is: (in http://www.dltk-kids.com/crafts/cartoons/spongesong.htm)

Are you ready kids?
Aye-aye Captain.
I can't hear you. ...
Aye-Aye Captain!!
Oh! Who lives in a pineapple under the sea?
SpongeBob SquarePants!
Absorbent and yellow and porous is he!
SpongeBob SquarePants!
If nautical nonsense be something you wish. ...
SpongeBob SquarePants!
Then drop on the deck and flop like a fish!
SpongeBob SquarePants!
Ready?

I work as a cook in Mr. Crab's restaurant, and I prepare the most delicious meals according to my great-grandmother's secret recipe book. These recipes have turned Mr. Crab's restaurant into an extremely famous gourmet place in contrast to Mr. Plankton's restaurant, which hardly ever sees any customers. Mr. Plankton and his wife Karen spend hours in front of the computer screen trying to discover a secret recipe which will attract all the customers. After trying really hard to find something that would cause trouble, he came up with a plan which has proved quite successful so far. He somehow managed to convince Pat the Pirate to take a few days off and go on a holiday and he called Lavinia the Maze to sing the rhyme every morning. Most of the citizens of Bikini Bottom didn't know that Lavinia is a witch, and she turned everything upside down when she changed the rhyme. So, she sings:

Aye-aye Lavi.
I can clearly hear you …
Aye-Aye Lavi!!
Oh! What does a pineapple do in the sea?
SpongeBob StonePants!
Dangerous, messy, and heavy is he!
SpongeBob StonePants!
If wonderful cities be something you wish. …
SpongeBob StonePants!
Then drop off the deck and sink like a wreck!
SpongeBob StonePants!
Ready?

The moment everybody sings the rhyme, everything in the city of Bikini Bottom changes, as all the items leave their place and some of them start to float while others sink! In this way the city has become a real mess, as nothing can stand still. You can see chairs float, cars sink, and even my pineapple house floats on the sea surface. I tried to solve the mystery by seeking the truth in my great-grandmother's books, and I think I eventually have a rescue plan. I found it in the phrase: "Should the city be turned upside down one day, seek for the real properties of water right away." This means that we have to find out which items float or sink when put in the water. If we draw or make a model city with these items, everything will be in place and Lavinia will go away. Everybody in the city desperately offered his help except for Mr. Plankton and his wife, of course, who try to mislead our logical thinking by playing tricks on the computer. It is extremely important to discover the scientific truth that will save our city. My dearest kids, your help will be invaluable.

With all my love,
SpongeBob.

APPENDIX B

Questionnaire

Name:
Surname:
Year of Study:
Student Number:
School:
Date:
Questions:
Part A: Extract from the questionnaire Views on Science-Technology-Society

(VOSTS) (Aikenhead, 1989).

Your position basically: (Please read the following statements and then choose one.)

1. Defining science is difficult because science is complex and does many things. But MAINLY science is:

A. A study of fields such as biology, chemistry, and physics.

B. A body of knowledge such as principles, laws, and theories that explain the world around us (matter, energy, and life).

C. Exploring the unknown and discovering new things about our world and universe and how they work.

D. Carrying out experiments to solve problems of interest about the world around us.

E. Inventing or designing things (for example, artificial hearts, computers, space vehicles).

F. Finding and using knowledge to make this world a better place to live in (for example, curing diseases, solving pollution, and improving agriculture).

G. An organization of people (called scientists) who have ideas and techniques for discovering new knowledge.

H. No one can define science.

I. I don't understand.

J. I don't know enough about this subject to make a choice.

K. None of these choices fit my basic viewpoint.

2. Some communities produce more scientists than other communities. This happens as a result of the upbringing which children receive from their family, schools, and community. Upbringing is mostly responsible:

A. Because some communities place greater emphasis on science than other communities.

B. Because some families encourage children to question and wonder. Families teach values that stick with you for the rest of your life.

C. Because some teachers or schools offer better science courses or encourage students to learn more than other teachers or schools.

D. Because the family, schools, and community all give children with an ability in science the encouragement and opportunity to become scientists.

E. It's difficult to tell. Upbringing has a definite effect, but so does the individual's qualities (for example, intelligence, ability, and a natural interest in science). It's about half and half.

Intelligence, ability and a natural interest in science are mostly responsible:

F. In determining who becomes a scientist. However, upbringing has an effect.

G. Because people are born with these traits.

H. I don't understand.

I. I don't know enough about this subject to make a choice.

J. None of these choices fit my basic viewpoint.

3. Science and technology influence our everyday science, and technology gives us new words and ideas.

A. Yes, because the more you learn about science and technology, the more your vocabulary increases and thus the more information you can apply to everyday problems.

B. Yes, because we use the products of science and technology (for example, computers, microwaves, health care). New products add new words to our vocabulary and change the way we think about everyday things.

C. Science and technology influence our everyday thinking BUT the influence is mostly from new ideas, inventions, and techniques which broaden our thinking.

Science and technology are the most powerful influences on our everyday lives, not because of words and ideas:

D. But because almost everything we do, and everything around us, has in some way been researched by science and technology.

E. But because science and technology have changed the way we live.

F.No, because our everyday thinking is mostly influenced by nonscientific things. Science and technology influence only a few of our ideas.

G. I don't understand.

H. I don't know enough about this subject to make a choice.

I. None of these choices fit my basic viewpoint.

4.   There seem to be two kinds of people: those who understand the sciences and those who understand the arts (for example, literature, history, business, law). But if everyone studied more science, then everyone would understand the sciences.

A. There ARE these two kinds of people. If the arts people did study more science, they would come to understand science, too, because the more you study something, the more you come to like and understand it.

There ARE these two kinds of people. If the arts people did study more science, they would not necessarily come to understand it better:

B. Because they may not have the skill or the talent to understand science. Studying will not give them this skill.

C. Because they may not be interested in science. Studying will not change their interest.

D. Because they may not be oriented or inclined towards science. Studying science will not change the kind of person you are.

E. There are not just two kinds of people. There are as many kinds as there are individual preferences, including people who understand both arts and sciences.

F. I don't understand.

G. I don't know enough about this subject to make a choice.

H. None of these choices fit my basic viewpoint.

5. Scientists have practically no family or social life because they need to be so deeply involved in their work.

A. Scientists need to be very deeply involved in their work in order to succeed. This deep involvement takes one away from one's family and social life.

B. It depends on the person. Some scientists are so involved in their work that their families and social lives suffer. But many scientists take time for family and social things.

C. At work scientists look at things differently than other people, but this doesn't mean they have practically no family and social lives.

A scientist's family and social lives are normal:

D. Otherwise their work would suffer. A social life is valuable to a scientist.

E. Because very few scientists get so wrapped up in their work that they ignore everything else.

F. I don't understand.

G. I don't know enough about this subject to make a choice.

H. None of these choices fit my basic viewpoint.

6.   Scientists trained in different countries have different ways of looking at a scientific problem. This means that a country's educational system or culture can influence the conclusions which scientists reach. The country DOES make a difference:

A. Because education and culture affect all aspects of life, including the training for thinking about a scientific problem.

B. Because each country has a different system of teaching science. The way scientists are taught to solve problems makes a difference in the conclusions scientists reach.

C. Because a country's government and industry will only fund science projects that meet their needs. This affects what a scientist will study.

D. It depends. The way a country trains its scientists might make a difference to some scientists. BUT other scientists look at problems in their own individual way based on personal views.

The country does NOT make a difference:

E. Because scientists look at problems in their own individual way regardless of what country they were trained in.

F. Because scientists all over the world use the same scientific method, which leads to similar conclusions.

G. I don't understand.

H. I don't know enough about this subject to make a choice.

I. None of these choices fit my basic viewpoint.

7. Scientific ideas develop from hypotheses to theories, and finally, if they are good enough, to being scientific laws.

Hypotheses can lead to theories, which can lead to laws:

A. Because a hypothesis is tested by experiments, if it proves correct, it becomes a theory. After a theory has been proven true many times by different people and has been around for a long time, it becomes a law.

B. Because a hypothesis is tested by experiments, if there is supporting evidence it's a theory. After a theory has been tested many times and seems to be essentially correct, it's good enough to become a law.

C. Because it's a logical way for scientific ideas to develop.

D. Theories can't become laws because they both are different types of ideas. Theories are based on scientific ideas which are less than 100% certain, and so theories can't be proven true. Laws, however, are based on facts only and are 100% sure.

E. Theories can't become laws because they both are different types of ideas. Laws describe things in general. Theories explain these laws. However, with supporting evidence, hypotheses may become theories (explanations) or laws (descriptions).

F. I don't understand.

G. I don't know enough about this subject to make a choice.

H. None of these choices fit my basic viewpoint.

8. When scientists investigate, it is said that they follow the scientific method. The scientific method is:

A. Controlling experimental variables carefully, leaving no room for interpretation.

B. A logical and widely accepted approach to problem solving.

C. Testing and retesting–proving something true or false in a valid way.

D. Postulating a theory, then creating an experiment to prove it.

E. Questioning, hypothesizing, collecting data, and concluding.

F. Considering what scientists actually do, there is no such thing as the scientific method.

G. I don't understand.

H. I don't know enough about this subject to make a choice.

I. None of these choices fit my basic viewpoint.

9. Scientists should NOT make errors in their work because these errors slow the advance of science.

A. Errors slow the advance of science. Misleading information can lead to false conclusions. If scientists don't immediately correct the errors in their results, then science is not advancing.

B. Errors slow the advance of science. New technology and equipment reduce errors by improving accuracy and so science will advance faster.

Errors CANNOT be avoided:

C. So scientists reduce errors by checking each others' results until agreement is reached.

D. Some errors can slow the advance of science, but other errors can lead to a new discovery or breakthrough. If scientists learn from their errors and correct them, science will advance.

E. Errors most often help the advance of science. Science advances by detecting and correcting the errors of the past.

F. I don't understand.

G. I don't know enough about this subject to make a choice.

H. None of these choices fit my basic viewpoint.

Part B: Which type of scientific method can you find in the following extracts of science activities or dialogues? Choose only one answer.

1. How will the ant be able to travel across the pond? Of the items it used to make a boat, which ones float and which ones sink?

    a. Communication
    b. Experiment
    c. Measurements
    d. Observation

2. Students are presented with a variety of materials that sink or float.They talk about the properties and the characteristics of each material by answering questions such as: What is this? In what way is it different from...?

    a. Testing predictions
    b. Classification using defined criteria
    c. Observation
    d. Predictions

3. Students discuss which materials float or sink.

    a. Concluding
    b. Predictions
    c. Testing predictions
    d. Classification using children's criteria

4. Students put vegetables in a bowl of water and observe which of them sink and which ones float.

    a. Experiment
    b. Prediction
    c. Testing predictions
    d. Classification using children's criteria

5.   Students blow their boats with a straw and see which of them travels faster.

    a. Communication
    b. Prediction
    c. Testing predictions
    d. Observation

6. Using their senses, students observe the materials and make predictions about what will happen if they put each material in the water.

    a. Experiment
    b. Prediction
    c. Testing predictions
    d. Classification using children's criteria

APPENDIX C

Interview

Name:
Surname:
Year of Study:
Student Number:
School:
Date:
Questions:

1. How did the pupils respond to the fact that they received a letter from a popular cartoon character asking for their help?
2. Did they know about SpongeBob?
3. How did they react to the teacher in role? Were they persuaded? Were they willing to participate?
4. Which was the dominating element that they chose to paint?
5. What happened during argumentation? In what way did they make contact with the materials?
6. Did the teacher intervene when the pupils filled in the board of predictions?
7. Did the presence of SpongeBob play any particular role during the experiment?
8. What were the cognitive obstacles that pupils faced connected with floating and sinking concepts.
9. What was the impact of the telephone conversation?
10. How did you, as a teacher, feel while you were in the role of SpongeBob?
11. Evaluate the use of cartoons for teaching floating and sinking concepts.
12. Which part of the didactic scenario do you consider the most significant?
13. Did you obtain any knowledge about teaching natural sciences? Did you connect it with other fields of knowledge?
14. Comment on the value of the whole procedure for designing natural science activities.
15. If you were to redo the whole procedure, would you make any changes?

REFERENCES

Abd- El Khalick, F., Bell, R., & Lederman, N. (1997). The nature of science and instructional practice: Making the unnatural natural. *Science Education, 82*, 417–436.

Aikenhead, G. S., & Ryan, A. G. (1989). *The development of a multiple-choice instrument for monitoring views on science-technology-society topics*. Ottawa: Social Sciences and Humanities, Research Council of Canada.

American Association for the Advancement of Science (AAAS). (1989). *Science for all Americans: Project 2061*. Washington, DC: AAAS.

Bob Sponge rhyme. Retrieved November 24, 2010, from http://www.dltk-kids.com/crafts/cartoons/spongesong.htm.

Bob Sponge Squarepants. Retrieved November 24, 2010, from http://www.spongebob-games.com/.

Bongco, M. (2000). *Reading comics: Language, culture and the concept of the superhero in comic books*. New York: Garland Publishing Inc.

Bravo-Adúriz, A. (2005). *An introduction to the nature of science*. Fondo de Cultura Económica, Buenos Aires [in Spanish].

British Cartoon Archive. Retrieved September 30, 2009, from http://www.cartoons.ac.uk/teachingaids.

Carter, J. (2008). Dora the explorer: Preschool geographic educator. *Journal of Geography, 107*(3), 77–86.

Cohen, L., & Manion, L. (1994). *Research methods in education* (4th ed.). London: Routledge.

Cole, M., & the Distributed Literacy Consortium. (2006). *The fifth dimension: An after-school program built on diversity.* New York: Russell Sage Foundation.

Comic Book Museum. Retrieved August 5, 2009, from http://www.nyccomicbookmuseum.org/education.htm.

Comic Life. Retrieved October 1, 2009, from http://plasq.com.

Cross-Thematic Curriculum Framework for Nursery School. (2003). Available at: http://www.pi-schools.gr.

Dewey, J. (1934). *Art as experience.* New York: Capricorn.

Eijck, van M., & Roth, W-M., (2010). Theorizing scientific literacy in the wild. *Educational Research Review, 5,* 184–194.

Engeström, Y. (1987). *Learning by expanding: An activity theoretical approach to developmental research.* Helsinki: Orienta-Kousultit.

Engeström, Y. (2005). *Developmental work research: Expanding activity theory in practice.* Berlin: Lehmanns Media.

Fleer, M., & Hedegaard, M. (2010). Children's development as participation in everyday practices across different institutions. *Mind, Culture and Activity, 17*(2), 149–168.

Gardner, H. (1983). *Frames of mind: The theory of multiple intelligences.* New York: Basic Books.

Hedegaard, M., & Fleer, M. (2008). *Studying children. A cultural-historical approach.* London: Open University Press.

Hogan, K. (2000). Exploring a process view of students' knowledge about the nature of science. *Science Education, 84,* 51–70.

Horn, M. (1980). *The world encyclopaedia of cartoons* (Vol. 1). New York: Chelsea House.

Keogh, B., & Naylor, S. (1999). Concept cartoons, teaching and learning in science: An initial evaluation. *Public understanding of Science, 8,* 1–18.

Laugksch, R., (2000). Scientific literacy: A conceptual overview. *Science Education, 84*(1), 71–94. John Wiley & Sons, Inc.

Lederman, N. (1992). Students' and teachers' conceptions of the nature of science: A review of the research. *Journal of Research in Science Teaching, 29,* 331–359.

Mayer, R. (1996). Learners as information processors: Legacies and limitations of educational psychology's second metaphor. *Education Psychology, 31,* 151–161.

McCloud, S. (1993). *Understanding comics: The invisible art.* Northampton, MA: Kitchen Sink Press.

McComas, W. (Ed.). (1998). *The nature of science in science education. Rationales and strategies.* Dordrecht: Kluwer.

McComas, W. (2005). *Teaching the nature of science: What illustrations and example exist in popular books on the subject?* Paper presented at the IHPST Conference, Leeds (UK), July 15–18.

Miller, J. D. (1983). Scientific literacy: A conceptual and empirical review. *Daedalus, 112*(2), 29–48.

Miller, J. D. (2007). *The public understanding of science in Europe and the United States.* A paper presented to the 2007 annual meeting of the American Association for the Advancement of Science, San Francisco.

National Association for the Education of Young Children (NAYEC). (2002). Conezio, K. &French. Science in the pre-school classroom. Capitalizing on Children's Fascination with the Everyday World to Foster Language and Literacy Development. Young Children, downloaded on March 15, 2009, from www.naeyc.org/resources/journal

National Research Council (NRC). (1996). *National science education standards.* Washington, DC: National Academy Press.

Plakitsi, K. (2007). *Didactic of natural sciences in pre-primary and primary childhood: Contemporary trends and prospective.* Athens: Patakis [in Greek].

Roth, W.-M., & Tobin, K. (2002). Redesigning an "Urban" teacher education program: An activity theory perspective. *Mind, Culture, and Activity, 9*(2), 108–131.

Roth, W.-M. (2005). *Doing qualitative research.* Rotterdam: Sense Publishers.

Roth, W.-M., & Lee, S. (2004). Science education as/for participation in the community. *Science Education, 88*(2), 263–291.

Unesco. (2007). *Literacy initiative for empowerment, LIFE 2006–2015*. Paris: Unesco.
Unesco. (2008). *The global literacy challenge*. Paris: Unesco.
Unesco. (2010). Summit on the Millennium Goals. Retrieved November 15, 2010, from http://www.un. org/millenniumgoals/.
Vygotsky, L. S. (1978). *Mind in society*. Cambridge, MA: Harvard University Press.
Yang, G. (2003). *Comics in education*. Retrieved December 12, 2008, from www.humblecomics.com/ comicsedu/index.htlm.

Eleni Kolokouri
School of Education
University of Ioannina
Greece

Katerina Plakitsi
School of Education
University of Ioannina
Greece

EFTYCHIA NANNI & KATERINA PLAKITSI

9. BIOLOGY EDUCATION IN ELEMENTARY SCHOOLS

How Do Students Learn about Plant Functions?

INTRODUCTION

This study reports part of a wider project inspired by sociocultural theory, which was designed to investigate and improve pre- and in-service teachers' training on topics related to living things. The main purpose of the whole project is to design and develop helpful teaching materials, supporting tools, and a teacher education curriculum for living things.

The project was influenced by the continuing and growing interest of global educational community in the teaching of natural sciences at all levels of education (AAAS, 1989; OECD, 2007; Osborne & Dillon, 2008). In 2004, the European Union issued a study entitled "Europe Needs More Scientists," thus placing a high priority on science education (EC, 2004).

Considering this priority, we investigate innovative processes and practices in pre- and in-service teachers' training about living things. Living things is a topic that most science curricula include in the very early years of schooling as foundational knowledge. However, research has shown that even in late elementary school there is a significant proportion of students who do not have an understanding of the concept of living things that corresponds with biological theory (Venville, 2004).

The theoretical framework is the cultural-historical activity theory (CHAT) that is inherently developed to understand the use of tools in human activity (Van Eijck & Roth, 2007). Learning in general, and science teacher education more specifically, can be seen as sites for social participation where cultural enactment occurs (Espinet & Ramos, 2009). The activity theorists argue that CHAT is a coherent theoretical framework which establishes science education as participation in the community (Roth & Lee, 2004). Moreover, as we posed in Chapter 1, CHAT bridges the gap between theory and praxis. Also, it aims to achieve the scope of interdisciplinary science education in multicultural Europe. Consequently, a new mentality has emerged which is concerned with situating science education as part of the society. This could reform science education from the inside in a natural and logical way, while lifelong learning activities take place in|for the community, which is inseparable from individuals (Plakitsi, 2009).

K. Plakitsi (ed.), Activity Theory in Formal and Informal Science Education, 229–251.

This study is a pilot research study which aims to investigate how primary school students conceptualize plant functions while students are involved with different systems of activities. The study promotes a bottom-up approach, which helps the young children's cognitive development concerning plant functions and also the development of school teachers' training programs in teaching science.

Based on the CHAT theoretical framework, we undertook a pilot implementation on teaching living things to a class of 16 primary students who were 10 to 11 years old (5th grade). We studied the way they conceptualize integrated concepts, for example, the system of living|non-living things and the system of photosynthesis|respiration|transpiration. We choose the dialectical symbol |, because we think of living at the same time as thinking of non living things. A living thing exists at one moment while it cannot exist at another and vice versa. So, we cannot conceptualize a living thing without simultaneously conceptualize a non living thing. On the other hand, photosynthesis and transpiration and respiration form a unit considered as plant functions. We failed in teaching those functions in isolation and, now, we test them with a systemic teaching approach. The teacher acts as researcher too, so we use the term teacher|researcher. The teacher was a female and she designed a sequence of tasks in order to help students to acquire learning about plant functions. Overall, our approach for the unit "Plants" was holistic, since our basic teaching objective is students coming to understand plants as living things with dynamic interactive and interdependent functions. So innovative teaching took place without using the current school textbooks on natural sciences, but using our own worksheets, which we had prepared for the purposes of the pilot study. Plants and their functions constitute a whole that cannot be divided into subunits (especially when the students are from 10 to 11 years old).

The analysis of the activities gave us a large number of answers, but also raised new research questions. Using CHAT for science education in the field of living things seems promising. In addition, this theoretical framework seems to be appropriate and fruitful for teachers|researchers' education in primary science teaching. The experimentation on new processes in natural science pedagogical actions in the wider system of activities demonstrates the importance of CHAT for science education, and encourages us to continue our studies in this research field. This case study, even though limited, could actuate the discussion of cultural studies of science education, especially teacher training for applications of the CHAT framework in science education.

RATIONALE

The rationale contains the general problem, the specific problem of teaching plant functions, the innovative teaching proposal following a didactical transformation, the multiple levels of expected outcomes and the necessity of choosing the CHAT framework.

General problem: In recent years, many countries have reformed their science curriculum. Major agencies stress the need for scientific literacy and develop programs for both evaluating the results of the teaching of science (Programme for International Student Assessment – PISA) and for the promotion of science (Project 2061).

For this purpose the Greek Ministry of Education changed the curriculum in 2003 towards Cross Curricular/Thematic Framework (CCTF) and the Individual Subject Curricula for compulsory education, which introduced a cross-thematic approach to learning. Educational change in Greece focused on the preservation of national identity and cultural heritage, on the one hand, and the development of European citizenship awareness, on the other. The final curriculum is the end product of a long, strenuous, and collaborative effort of teachers, educational administrators, and scientific societies with the co-ordination of the Greek Pedagogical Institute. Currently, this book's editor, Katerina Plakitsi is co-ordinating an ongoing reform towards a new curriculum for the new school of the 21st century. This new curriculum is now in its pilot implementation and testing. Our pilot research dovetails the two curricula enhancing inquiry based learning and the cultural dimension of science education.

In both curricula, the teaching of natural sciences in compulsory education in Greece and especially in primary schools is integrated (including physics, chemistry, biology, geology, and geography) into the subject called "Studies of the Environment," taught in the first four grades, and in the subject called "Explore the Natural World," taught in the last two grades. One exception is geography, which is taught as an independent subject in the last two grades of primary school.

According to the new pilot curriculum for Natural Sciences, the aim of teaching science in compulsory education is incorporated in the general aims of education, which are the well-rounded and balanced development of the individual through the development of critical thinking abilities and a positive attitude towards creative action on a personal and a social level. In our approach we try to move the discussion more from the society to the individual and less from the individual to the society. The new curriculum for Natural Sciences sets as priorities:
- the respect for cultural diversity and gender equality
- the promotion of personal and social recognition
- The motivation of the student towards democratic and civic participation.

The innovation of the new curriculum for natural sciences is at multiple levels. The main innovative aspects of this curriculum are:
- the networking of concepts,
- the attempt to create authentic learning environments,
- the attempt of cultivate the language and especially the argument,
- the absorbtion of Information and Communication Technologies (ICT) as an integral part of teaching science,
- the opening of science to society and culture,
- the cultivation of elements from the Nature of Science,
- the development of skills from the world of Science and Technology and their transformation into abilities for the modern citizen.

Especially, the new pilot curriculum includes a major thematic unit called "The life around us" where the module of living things is dominant in all grades of primary school.

The specific problem of teaching plant functions: The module of living things is a main subject of biology which is a subfield of natural sciences and concerns the study of life in living organisms such as plants, animals, fungi, protista, bacteria, and all other forms of life. Most science curricula include it in the very early years of schooling as foundational knowledge. However, research has shown that even in late elementary school there is a significant proportion of students who do not have an understanding of the concept of living things that corresponds with biological theory (Venville, 2004). That the biology course includes lots of abstract concepts may cause students to have difficulty in constructing knowledge (Keles and Kefeli, 2010). In our opinion, we may have failed because we teach living things in unauthentic learning environments and often withoutt any enactment of learning communities.

Moreover, this pilot researching study faces the problem of inadequate teaching of plant functions. Plant functions are a difficult biological topic for the students, partially because they are characterized by a number of conceptions that usually make the teaching of these concepts inadequate (Canal, 1999; Carlsson, 2002; Keles & Kefeli, 2010; Lumpe & Staver, 1995; Marmaroti & Galanopoulou, 2006; Yelminez & Tekkaya, 2006).

The concepts of respiration and photosynthesis are important in the wider context of ecological understanding. For example, Lumpe and Staver (1995) report that for students to begin to understand life and life processes, they must understand the concepts associated with plant nutrition. Biologists and biology educators state that understanding energy issues in organisms, namely respiration and photosynthesis, is key to understanding more global issues, such as energy flow, food supplies, and other ecological principles. Nevertheless, during her research related to ecological understanding Carlsson states (2002), in summary, that photosynthesis is understood in many different ways and that the process of respiration in general, and in plants in particular, is unknown to a majority of students.

In depth, and according to a study by Yelminez and Tekkaya (2006), students' conceptualizations of photosynthesis and respiration in plants are:

1. Carbon dioxide is used in respiration, which only occurs in green plants when there is no light energy to photosynthesize.
2. Respiration in plants takes place in the cells of the leaves only.
3. Respiration is the exchange of carbon dioxide and oxygen gases through plant stomata.
4. Green plants take in carbon dioxide and give off oxygen when they respire.
5. Green plants respire only at night, when there is no light energy.
6. Green plants do not respire; they only photosynthesize.
7. Photosynthesis provides energy for plant growth.
8. Plants respire when they cannot obtain enough energy from photosynthesis, and animals respire continuously because they cannot photosynthesize.

Canal (1999) argues that the common conceptions in primary school students regarding green plant nutrition are:

1. Plant nutrition is the process through which plants feed themselves.
2. Plants are nourished by the substances that they take from the earth through their roots: water and mineral substances.
3. The substances that plants absorb from the earth form raw sap, which moves through the stalk and which allows the plant to grow and carry out its other vital functions.
4. Sunlight is indispensable for the health of green plants, their strength, and good color. Without light, they become weak and may end up dying.
5. Plants breathe like animals, taking in and expelling air. If they do not do so continuously, they asphyxiate and die, like people.

The innovative teaching proposal following a didactical transformation: This pilot study attempts an innovative teaching proposal concerned with the didactical content matter and investigates the conceptualizations that children make about the processes of plants (photosynthesis, respiration, transpiration). For this purpose we proposed a unified topic of plants processes and taught it to primary school students of 10 to 11 years old.

Our teaching proposal tries to integrate the concepts of photosynthesis – respiration – transpiration in a common context and in this way overcoming the students' difficulties in learning these concepts. These concepts are dynamic interactive and interdependent processes, and not separate processes, which means that each process interacts with each other and depends on the others.

More specifically, plants, as living organisms, respire and by this process obtain oxygen and remove carbon dioxide. Photosynthesis is the process of converting light energy to chemical energy and storing it in the bonds of sugar. Respiration is the process where cells use this food (sugar) to release stored energy. So, respiration and photosynthesis are closely related and opposite processes. The complex substances (sugar) which formed during photosynthesis are broken down into simpler ones in a process which uses oxygen (respiration). On the other hand, water in the roots is pulled through the plant by transpiration (loss of water vapor through the stomata of the leaves). Transpiration uses about 90% of the water that enters the plant. The other 10% is an ingredient in photosynthesis and cell growth. Without the process of transpiration, plants would not be able to complete their food production or photosynthesis.

This study sees the plant processes in the context of living systems and supports the connection of elementary science to everyday life. Despite the many difficulties students have in understanding the concepts of plant processes, these processes play important roles in the understanding of many aspects of living systems. Also, one of the aims of science education is to make students learn meaningfully and use their learning to satisfy their needs in daily life. In this sense, considering all above studies and researchers, we designed this pilot study to achieve two broad goals: (1) examine students' conceptualization about living things and (2) develop school teachers' training programs in teaching science.

The multiple levels of expected outcomes: Our innovative teaching proposal traces new paths in elementary science education and meets the cultural – historical activity theory framework. It is well-known that every science has its own epistemology, methods, methodology and consequently its own specificities. In this sense, science education is a field with its own specificities that are not found in other fields. As Katerina Plakitsi mentioned in Chapter 3, typically science education takes place in a school and is structured by a curriculum (formal science education). At the same time, science education could include out-of-school sites, such as science centres, museums, media, community-based programs, and new digital learning environments (informal science education). In any case, science education comprises science content, teaching pedagogy and social science.

As stated in previous chapters, according to the curricula for natural sciences, teaching science has some general aims on a personal and social level, which are well-connected with scientific knowledge and scientific literacy. But science education refers not only to the science content, but also to pedagogical and social practices. Teaching and learning science occurs in a social-cultural context and so we do not only focus on science content. This aspect can be more obvious, through the expected outcomes of science education.

Below we pose the expected outcomes of science education, focused on the topic of plant functions that we studied in our pilot research.

Students of 10 to 11 years old should:
– Be able to identify the parts of a plant.
– Explain in simple terms the function of photosynthesis.
– Find out experimentally that plants breathe.
– Explain that the functions of photosynthesis and respiration are contrary processes.
– Find out experimentally the function of transpiration.

– Feel a sense of satisfaction brought about by being engaged with science and natural phenomena, such as plant functions.
– Develop positive attitudes and values about life and living things.
– Develop interest in exploring nature.
– Acquire motivation to learn science.

– Develop skills in arranging experiments.
– Develop creativity through active participation.
– Develop communication skills, such as non-discursive communication (body language, facial expressions, etc.).

The necessity of choosing the CHAT framework: All above specificities of science education gave us the impetus to study all the interactions taking place during the teaching through the theoretical framework of CHAT. CHAT gives this appropriate context to study these interactions and to analyse most aspects of science education.

Also, given that there is much research on science education, only a few studies use CHAT. The activity theorists argue that CHAT is a coherent theoretical framework which establishes science education as participation in the community (Roth and Lee, 2004). It aims to achieve the scope of interdisciplinary science education in multicultural Europe and situates science education as part of the society.

METHODS

This exploratory research used a case study design and qualitative data collection methods to investigate the process of learning when 10-year-old students (5th grade) are learning about plants and their functions in the context of CHAT. We studied the way they conceptualize integrated concepts of plant functions (photosynthesis-respiration-transpiration). The research methodology followed a qualitative design, involving classroom observations and content analysis. We collected data by videorecordings and analyzed how biology education is progressing at schools under the CHAT framework. Tobin (2006) argues that research in classrooms focuses on better understanding of teaching and learning and using what is learned to create and sustain improved learning environments.

The sample used in this pilot study consisted of 16 students in primary school (5th grade) from the island of Corfu. All of them were in mixed-ability groups and had not received any instruction about plant functions in previous lessons on natural sciences.

Eftychia, acting as teacher|researcher has 4 years of teaching experience in the early years of schooling. She participated in the study anticipating that the expected feedback about children's conceptualizations of plant functions can be used to plan future teaching approaches. The research consisted of 3 days of out-of-school in-service training of the teacher|researcher, 4 days in class preparation of the experiments, and 1 day in-class teaching. The 16 students were from mixed cultural and socioeconomic backgrounds, including children whose parents originated from Greece, Albania, and the Czech Republic. The teacher|researcher informed parents and guardians of all children in the class about the research and informed them of their right to withdraw their child from participation in the research at any time. No parent or guardian chose to withdraw a child from this research. Students also were informed in simple terms about the reasons for the researcher's presence in the classroom and were given the choice of whether they wanted to participate or not. No student decided to withdraw from the research.

According to the Greek National Curriculum, natural sciences lessons in 5th and 6th grades of primary school should take place three times a week (3 different teaching hours). But the school science textbooks keep an old and traditional module of plants consisting of four separate lessons, which are taught during four different teaching hours. The four lessons are (a) The Parts of Plants, (b) Photosynthesis, (c) Respiration, and (d) Transpiration. Our teaching approach was partly based on the above lessons (as we kept the same experiments), but took place in a different articulation. Our approach for the unit "Plants" was holistic, since our basic teaching objective is students coming to understand plants as living things with dynamic interactive and interdependent functions. So the innovative implementation took place without using the current school science textbooks, but using our own worksheets, which we had prepared for the purposes of research (see Appendixes I, II, III). The innovative implementation lasted almost 1 school day (4 teaching hours), since we wanted to approach the unit "Plants" within the same day and not to teach the three plant functions separately on different days. Plants and their functions constitute a whole, which cannot be divided into subunits (especially for students aged 10 to 11 years old).

The innovative implementation consisted of two different parts: one part of preparation and one part of class teaching. Both parts were done with the cooperation of the teacher|researcher and students. Also, the teacher|researcher designed the instruction according to the interests of students. For 4 days (Monday-Thursday) students and teacher|researcher collected the materials for the experiments and prepared the experiments, since most of them needed more than 1 day to be prepared (see Figures 1 and 2). For example, they had to cover some leaves of a plant with foil for 3 days, before studying what happens to leaves without light (photosynthesis).

The teaching of plant functions occurred on Friday, when all experiments were ready. That day was named by students "The Flower Day," since they were very excited about the expected teaching. The main teaching strategy was the creation of cooperative groups of students and the study of cooperative learning according to the CHAT framework. The teacher|researcher divided students into four groups (four students each), using a cooperative approach for each group. The research methodology followed a qualitative design, involving classroom observations and content analysis.

Figure 1. Students are preparing the experiment for respiration 3 days before the lesson (photo from author's archive).

Figure 2. Students are preparing the experiment for transpiration 4 days before the lesson (photo from author's archive).

The teacher|researcher designed a series of learning tasks so that students could approach inquiry based learning about plant functions. At first, teaching plant functions started with a review of plant parts and what plants need to grow. All students had some prior knowledge of both of these concepts. Each student was asked to say his/her opinion about the plants' needs for growth. Most students answered "sunshine, water, and soil." The teacher|researcher introduced students to the new concepts by presenting them with a computer-aided presentation about the three main plant functions. The presentation contained mainly shapes and images, since the new terms were very difficult for most children to understand (see Figures 3 and 4). According to the Mauseth (2009), the definitions of these three concepts of biology education are:

Photosynthesis is the combination of carbon dioxide with water to form carbohydrates in the presence of light.

Respiration is the process that breaks down complex carbon compounds into simpler molecules and simultaneously generates the ATP used to power other metabolic processes.

Transpiration causes leaf cells to lose water.

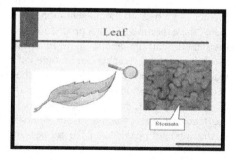

Figure 3. Slide presentation for the structure of leaf (photo from author's archive).

237

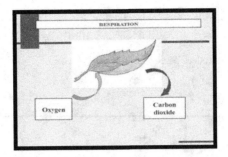

Figure 4. Slide presentation for the plant respiration (photo from author's archive).

After teacher|researcher's presentation the, students moved on to the next actions: the experimental testing and implementation of the new learning concepts. These actions took place through a series of teaching strategies, mainly through collaborative work between students and teacher during the experiments. For this reason, the teacher|researcher gave the students a worksheet (one for each group), in which they had to work and do the experiments (see Appendix I). The sequence of experiments was as follows (the items below are outcomes but not the experiments themselves.)

1. Finding out that humans expire carbon dioxide when they breathe out.
2. Discovering that plants also exhale carbon dioxide.
3. Finding out that certain foods contain starch.
4. Discovering that leaves contain starch also.
5. Discovering that leaves turn yellow if there is no light.
6. Discovering that yellowed leaves contain no starch.
7. Discovering that plants lose water into the atmosphere.
8. Discovering that plants lose water through their leaves.
9. Discovering that plants lose water mainly through the lower surfaces of their leaves.

All experiments involved observations, collecting data by students, who worked in groups as they conducted the experiments. During the experiments, very often students asked their teacher to go back to the presentation to discuss some concepts or to check their results of the experiments. This procedure gave students a positive feedback and supported their learning about plant functions. This was mainly achieved through discussion among teacher and students and among groups. Simultaneously, the teacher|researcher gave the groups another two worksheets, on which the students had to check their understanding about the new concepts (see Appendixes II and III). In this way the teacher|researcher evaluated what students had learned and if the teaching was effective. Both worksheets were of varying levels of difficulty for students and necessitated the cooperation of group members and the development of critical thinking. Open-ended questions and problems from everyday life were selected, instead of multiple-choice or closed-ended questions.

Students had to compose their thoughts and what they had learned in order to answer the questions. For example, in worksheet B students had the opportunity to describe the main functions of a plant in their own way (by using shapes, images, words, etc.). All groups had a good understanding of plant functions and responded positively to their evaluation.

ANALYSIS OF ACTIVITIES

The research rationality led us to apply CHAT framework as an appropriate framework to analyse the activities which took place during our teaching. The use of the CHAT framework enables participants to provide a detailed description of the connections within the lesson, as teacher|researcher and students work together to achieve individual and common goals. The implementation of the CHAT context through teaching in a school classroom supports the concept that the community is continually changing, shaped by the interactions of the subjects. The CHAT model is represented by Engeström (1987) in the form of a triangle, where the subject interacts with the community, rules, division of labor, tools and object (artifact) to reach the outcome This model has become a common framework for representing the understandings based on this theory, as it provides a complete analysis for the structure of a given activity system, or the central activity. However, the third generation of activity theory needs to develop conceptual tools to understand dialogue, multiple perspectives, and networks of interacting activity systems (Engeström, 2001). In this study, the central activity is the educational system, which includes other interacting activity systems and consists of actions.

According to Leontyev (Leontyev, 1981) there is a distinction between activities and actions. Activity (in human beings) is governed by its motive/motives and satisfies a need. Activity is carried out by a community. Actions are governed by their goals and often make sense only in a social context of a shared work activity. Action is carried out by an individual or group. Leontyev argue that:

> Let us now examine the fundamental structure of human activity in the conditions of a collective labour process [...] Processes, the object and motive of which do not coincide with one another, we shall call 'actions.' We can say, for example, that the beater's activity is the hunt, and the frightening of game his action. (Leontyev, 1981)

For Engeström (1987) activity is a collective, systemic formation that has a complex mediational structure. An activity system produces actions and is realized by means of actions. However, activity is not reducible to actions. Actions are relatively short-lived and have a temporally clear-cut beginning and end. Activity systems evolve over lengthy periods of socio-historical time, often taking the form of institutions and organizations.

Also, the theorists argue that:

> The theory of activity proposes three levels of interaction between an organism and its environment. At the level of concrete operations, the critical factor in interactions is the external conditions involved in performing

operations. This level is prominent in the psyches of animals and unconscious operations in humans. The second level is the level of higher primates and humans. At this level the critical factors are the goals of actions and the means of performing the actions. The third level of interaction is the level of the individual personality and the complex social environment. The crucial components here are activity and the individual's motives for engaging in activity. (Grigorenko et al., 1997)

In this context, our central activity of the educational system consists of subjects (students, teachers, educational policy makers, parents), who act on certain objects, which are necessary in order to achieve the desired outcomes of education. The objects include knowledge, skills, values and attitudes students should exhibit that reflect the outcomes. The outcomes of education include the development of individual morally, intellectually, physically, socially and aesthetically. These outcomes are needed by both individual and society. This activity is mediated by tools, such as language, thought, schools, and curricula. The activity is also mediated by the community in which the activity is being carried out. The community provide the rules that regulate the activity and constrain actions and interactions within the activity system. Such rules may be the practices of the educational system and the collaboration between the subjects. Finally, the division of labor refers to the division of tasks between the members of the community, such as the responsibility that subjects share with the community in the activity system.

During the same time, the activity of the educational system interacts with other activity systems and consists of actions. One central interacting activity system is the activity system of teacher training. The improvement of pre- and in-service teachers' training is our main goal of our study, under which we investigated innovative processes and practices in the teaching of living things.

In this context, our innovative teaching of plant processes constitutes an action of the central activity system of the educational system and contains interacting actions that share a common goal. In Figure 7, we represented the interacting actions of the teaching of plant processes and the way they share the common goal. The interacting actions contain the action of teaching the concept of respiration, the action of teaching the concept of photosynthesis and the action of teaching the concept of transpiration. The teaching of plant processes aimed to the understanding of plants as whole entities. The main goal of the teaching was students learning about plants and their processes and the achievement of the outcomes (attitudes and values about life and living things, scientific literacy, motivation to learn science, communication skills). First of all, students learn the parts of a plant and their usefulness. They learn that plants are living organisms and therefore respire and then that plants are photosynthetic, which means that they manufacture their own food molecules using energy obtained from light. Also, students had to learn that plants lose water through the process of transpiration. All these concepts were the basis of understanding plants as living things with dynamic interactive and interdependent processes, and not separate processes. Our innovative approach to teaching, as mentioned in previous pages, followed a

240

specific transformation which fostered the understanding of concepts in an interdependent context.

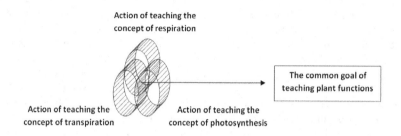

Figure 7. Interacting actions in the teaching of plant processes.

A Model of Achievement: The Goal of Plant Respiration

In this model, there are two interacting actions that are relevant to the teaching of the concept of plant respiration. The action of teaching that humans expel carbon dioxide when breathing out and the action of teaching that plants exhale carbon dioxide are foregrounded as constituting the understanding that plants breathe. These actions are inherently interrelated, in order to facilitate the understanding that plants breathe and absorb free molecules of oxygen (O_2) and use them to create water, carbon dioxide, and energy, which help the plants grow. In both actions, the subjects are the elementary students and the teacher, who interact with the community (school classroom), the rules as a basis for mediation of interactions (how a student collaborates with another student, a student with the teacher, a group with another group, a group with the teacher), and the division of labor (group work). The tools used for the action of teaching that humans expire carbon dioxide when breathing out were language, the body, thought, a worksheet, a glass, water, and a drinking straw. The tools used for the action of teaching that plants exhale carbon dioxide were language, the body, thought, a worksheet, two bottles, water, and celery. Both actions share a common goal, the understanding that plants breathe. The achievement of the common goal was made by the following process: Goal 1 of the first action was the human breath and carbon dioxide. Goal 1 moves from an initial state of an "experiment finding" to a collectively meaningful goal 2 constructed by the pedagogical action (the understanding of presence of carbon dioxide in human breath). Goal 1 of the second action was the plants and carbon dioxide. Also, this goal 1 moves to a collectively meaningful goal 2 constructed by the pedagogical action (the understanding of the presence of carbon dioxide around living plants). Both goals move to potentially shared or jointly constructed goal 3, which is the understanding that plants breathe.

A Model of Achievement: The Goal of Photosynthesis

In this model, there are four interacting actions that are relevant to the teaching of the concept of photosynthesis. The action of teaching that certain foods contain starch, the action of teaching that leaves contain starch, the action of teaching that leaves turn yellow if there is no light, and the action of teaching that yellowed leaves contain no starch are foregrounded as constituting the understanding that plants make their own food. These actions are inherently interrelated, in order to facilitate the understanding that plants make their own food, through photosynthesis, the process a plant uses to combine sunlight, water, and carbon dioxide to produce oxygen and sugar (energy). In all actions, the subjects are the elementary students and the teacher, who interact with the community (school classroom), the rules as a basis for mediation of interactions (how a student collaborates with another student, a student with the teacher, a group with another group, a group with the teacher), and the division of labor (group work). The tools used for the action of teaching that certain foods contain starch were language, thought, a worksheet, bread, potatoes, and iodine. The tools used for the action of teaching that leaves contain starch were language, thought, a worksheet, plant leaves, and iodine. The tools used for the action of teaching that leaves turn yellow if there is no light were language, thought, a worksheet, plants, and aluminum foil. The tools used for the action of teaching that yellowed leaves contain no starch were language, thought, a worksheet, plant leaves, and iodine. All actions share a common goal, the understanding that plants make their own food. The achievement of the common goal was made by the following process: Goal 1 of the first action was the concept "foods and starch." Goal 1 moves from an initial state of an "experiment finding" to a collectively meaningful goal 2 constructed by the pedagogical action (the understanding of the presence of starch in certain foods). Goal 1 of the second action was the concept of "plants and starch." Also, goal 1 moves to a collectively meaningful goal 2 constructed by the pedagogical action (the understanding of presence of starch in plant leaves). Goal 1 of the third action was the concept of "plants and sunlight," which moves from an initial state of an "experiment finding" to a collectively meaningful goal 2 constructed by the pedagogical action (the knowledge that plants need sunlight). Goal 1 of the fourth action was the concept of "absence of sunlight and sunlight." Also, this goal moves to a collectively meaningful goal 2 constructed by the pedagogical action (the knowledge that yellowed leaves – because of the absence of sunlight – contain no starch). All the goals move to potentially shared or jointly constructed goal 3, which is the understanding that plants make their own food.

A Model of Achievement: The Goal of Transpiration

In this model, there are three interacting actions that are relevant to teaching the concept of transpiration. The action of teaching that plants lose water into the atmosphere, the action of teaching that plants lose water through their leaves, and the action of teaching that plants lose water mainly through the lower surface of their leaves are foregrounded as constituting the understanding that plants lose water into the atmosphere. These actions are inherently interrelated, in order to facilitate the understanding that plants lose water to the atmosphere through small openings on the underside of leaves called stomata. In all actions, the subjects are the elementary students and the teacher who interact with the community (school classroom), the rules as a basis for mediation of interactions (how a student collaborates with another student, a student with the teacher, a group with another group, a group with the teacher), and the division of labor (group work). The tools used for the action of teaching that plants lose water into the atmosphere were language, thought, a worksheet, a plant, a transparent bag, and water. The tools used for the action of teaching that plants lose water mainly through their leaves were language, thought, a worksheet, plant leaves, bottles, plant shoots, and water. The tools used for the action of teaching that plants lose water mainly through the lower surface of their leaves were language, thought, a worksheet, plant leaves, water, plant shoots, and vaseline. All actions share a common goal, the understanding that plants lose water into the atmosphere. The achievement of the common goal was made by the following process: Goal 1 of the first action was the concept "plants and loss of water." Goal 1 moves from an initial state of an "experiment finding" to a collectively meaningful goal 2 constructed by the pedagogical action (the knowledge that plants lose water into the atmosphere). Goal 1 of the second action was the concept of "leaves and loss of water." Also, this goal 1 moves to the collectively meaningful goal 2 constructed by the pedagogical action (the knowledge that plants lose water mainly through their leaves). Finally, goal 1 of the third action was the concept of "loss of water and parts of leaves." Also, goal 1 moves from an initial state of an "experiment finding" to a collectively meaningful goal 2 constructed by the pedagogical action (the knowledge that plants lose water mainly through the lower surface of their leaves). All the goals move to potentially shared or jointly constructed goal 3, which is the understanding that plants lose water into the atmosphere.

CONCLUSIONS

The study described in this paper contributes to a limited but growing interest in CHAT-based education research. This research, even though limited, could give impetus to the discussion of cultural studies of science education and, especially, teacher training in applications of the CHAT framework in science education. The experimentation on new processes in natural science pedagogical actions in the wider systems of activities demonstrates the importance of CHAT for science education and encourages us to study this research field. We believe that this field could yield new and useful insights into formal and informal science education.

The results of this analysis provided many answers, but also raised new research questions. For example, our analysis has been used to examine the actions of teaching through helping students in elementary school learn about "plant functions." The interactive activity systems described in this study capture a moment in time of the learning process. The components of the CHAT triangle (community, subjects, etc.) are constantly changing as students evolve their thinking, explore new options, and interact with other activity systems. These issues should be further explored in order for CHAT to become better understood in educational research. Through this process, CHAT-based research and analysis should lead us to new forms of learning, knowing, and interacting.

Nevertheless, the results make us very optimistic about the implementation of a CHAT framework in biology education in elementary schools. Students participated actively in all educational activities and successfully achieved the object of the main activity system of teaching plant functions, as well the objects of the sub-activity systems. Also, students developed positive attitudes and values about living things and reinforced their creativity and communication skills through collaborative work. Doubtless, using CHAT in science education about living things seems promising. It might be a way to fulfill the main objectives of science education, which are the motivation of students to experience meaningful learning, the achievement of scientific literacy, and the adoption of social attitudes and values.

APPENDIX I

Plant functions: Worksheet A (Group ...)

When you blow into the glass containing limewater...

_____
_____
_____
_____
_____
_____
_____
_____

Words to use: Carbon dioxide, expire, inhale, oxygen, cloudy

You fill two bottles with limewater (place one stalk of celery in the first bottle) …

Words to use: respiration, oxygen, carbon dioxide

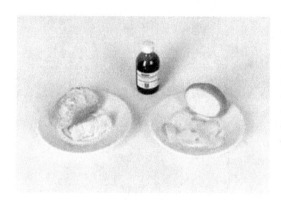

When you add iodine …

Words to use: starch, bread, potatoes, color

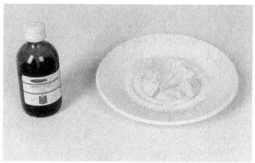

When you add iodine …

Words to use: discolored leaf, starch

245

You covered some leaves with aluminum foil …

Words to use: observe, color, yellow, light

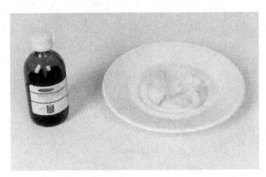

The leaves you had covered with foil …

Words to use: discolored leaf, iodine, light

You covered a plant in a pot with a transparent bag …

Words to use: water, light, drops

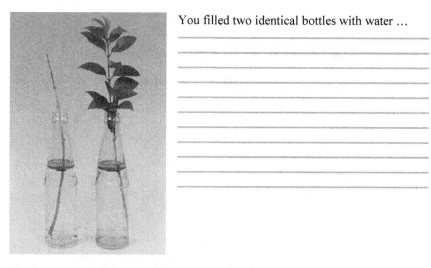

You filled two identical bottles with water …

Words to use: oil, light, stem, leaves, water level

You put similar stems of a plant in four different bottles …

Words to use: light, oil, stems, leaves, Vaseline, upper surface, lower surface, water level

APPENDIX II

Plant Parts and Their Processes: Worksheet B (Group ...)

Describe the main plant processes. You can use shapes, images, words, etc. See pictures below.

APPENDIX III

Plant pro: Worksheet C (Group ...)

A. Mary is home again after summer holidays. She opened the door of her room and went inside. Immediately she opened the shutters to get light into the room. She was very surprised when she saw that her favorite basil had wilted. But how did it happen? My grandfather watered it every day, thought Mary. Without wasting any time, she found her school textbook, which talked about plants. In one chapter, she found the plant experiments they had done in school. Unfortunately, the conclusions were missing. Can you help Mary understand what had happened to her basil?

B. Mike went on vacation for a few days and left this note to his friend Eva. Do you agree with the advice? Is it enough?

> Dear Eva,
> please take care of my dog and my plants:
> Put enough food and water for my dog.
> Keep the shutters open.
> Water the plants.
>
> Best wishes,
> Mike

C. "Respiration and photosynthesis are opposite processes." Can you explain this statement?

REFERENCES

American Association for the Advancement of Science (AAAS). (1989). *Science for all Americans.* New York: Oxford University Press.

Canal, P. (1999). Photosynthesis and 'inverse respiration' in plants: An inevitable misconception? *International Journal of Science Education, 21*(4), 363–371.

Carlsson, Br. (2002). Ecological understanding 1: Ways of experiencing photosynthesis. *International Journal of Science Education, 24*(7), 681–699.

Engeström, Y. (1987). *Learning by Expanding. An Activity-Theoretical Approach to Developmental Research.* Helsinki: Orienta-Konsultit.

Engeström, Y. (1999). Innovative learning in work teams: Analyzing cycles of knowledge creation in practice. In Y. Engeström, R. Miettinen, & R.-L. Punamaki (Eds.), *Perspectives on activity theory* (pp. 377–404). Cambridge: Cambridge University Press.

Engeström, Y. (2001). Expansive learning at work: Toward an activity theoretical reconceptualization. *Journal of Education and Work, 14*(1), 133–156.

Espinet, M., & Ramos, L. (2009, August 31–September 4). *Multilingual science education contexts: Opportunities for pre-service science teacher learning.* Paper presented at the symposium Cultural Studies of Science Education in Europe: Mapping Issues and Trends at the European Science Education Research Association (ESERA) biannual conference, Istanbul, Turkey.

European Commission. (2004). *Europe needs more scientists.* Report by the High Level Group on Increasing Human Resources for Science and Technology in Europe. Brussels: European Commission.

Irez, S. (2006). Are we prepared? An assessment of preservice science teacher educators' beliefs about nature of science. *Science Education, 90*(6), 1113–1143.

Keles, E., & Kefeli, P. (2010). Determination of student misconceptions in "photosynthesis and respiration" unit and correcting them with the help of CAI material. *Procedia Social and Behavioral Sciences, 2*, 3111–3118.

Krall, R., Lott, K., & Wymer, C. (2009). Inservice elementary and middle school teachers' conceptions of photosynthesis and respiration. *Journal of Science Teacher Education, 20*, 41–55.

Lin, C., & Hu, R. (2003). Students' understanding of energy flow and matter cycling in the context of the food chain, photosynthesis, and respiration. *International Journal of Science Education, 25*(12), 1529–1544.

Lumpe, A., & Staver, J. (1995). Peer collaboration and concept development: Learning about photosynthesis. *Journal of Research in Science Teaching, 32*(1), 71–98.

Marmaroti, P., & Galanopoulou, D. (2006). Pupils' understanding of photosynthesis: A questionnaire for the simultaneous assessment of all aspects. *International Journal of Science Education, 28*(4), 383–403.

Mauseth, J. (2009*). Botany. An introduction to plant biology.* Canada: Jones and Bartlett Publishers.

Organisation for Economic Co-operation and Development (OECD). (2007). *PISA 2006: Science competencies for tomorrow's world.* Volume 1: Analysis.Paris: OECD. Retrieved August 4, 2011 from http://www.keepeek.com/Digital-Asset-Management/oecd/education/pisa-2006_9789264040014-en.

Osborne, J., Dillon, J., & King's College London. (2008). *Science education in Europe: Critical reflections.* London: The Nuffield Foundation.

Palmer, D. (2008). Practices and innovations in Australian science teacher education programs. *Research in Science Education, 38*, 167–188.

Pedagogical Institute (PI). (2003). *Cross-thematic curriculum framework for natural sciences.* Translated from the official Gazette issue B, nr 303/13-03-03 and issue B, nr 304/13-03-03. Athens: PI. Retrieved August 4, 2011 from http://www.pi-schools.gr/download/programs/depps/english/19th.pdf.

Pedagogical Institute (PI). (2011). *Curriculum for natural science.* Retrieved October 8, 2011 from http://digitalschool.minedu.gov.gr/info/newps.php.

Plakitsi, K. (2009). *Activity theory in formal and informal science education (ATFISE project)*. Paper presented at the symposium Cultural Studies of Science Education in Europe: Mapping Issues and Trends at the European Science Education Research Association (ESERA) biannual conference, Istanbul, Turkey, August 31–September 4.

Reiss, M., & Tunnicliffe, S. (1999). *Building a model of the environment: How do children see plant?* Paper for NARST, Boston, March 28–31.

Roth, W.-M., & Lee, S. (2004). Science education as/for participation in the community. *Science Education, 88*(2), 263–291.

Smith, E., & Anderson, C. (1984). Plants as producers: A case study of elementary science teaching. *Journal of Research in Science Teaching, 21*(7), 685–698.

Tobin, K. (2006). Qualitative research in classrooms. In K. Tobin & J. L. Kincheloe (Eds.), *Doing educational research* (pp. 15–57). Rotterdam: Sense Publishers.

Van Eijck, M., & Roth, W. M. (2007). Rethinking the role of information technology-based research tools in students' development of scientific literacy. *Journal of Science Education and Technology, 16*(3), 225–238.

Venville, G. (2004). Young children learning about living things: A case study of conceptual change from ontological and social perspectives. *Journal of Research in Science Teaching, 41*(5), 449–480.

Uno, G. (2009). Botanical literacy: What and how should students learn about plants? *American Journal of Botany, 96*, 1753–1759.

Yenilmez, A., & Tekkaya, C. (2006). Enhancing students' understanding of photosynthesis and respiration in plant through conceptual change approach. *Journal of Science Education and Technology, 15*(1), 81–87.

Eftychia Nanni
School of Education
University of Ioannina
Greece

Katerina Plakitsi
School of Education
University of Ioannina
Greece

LIST OF CONTRIBUTORS

Eleni Kolokouri is a pre-primary school teacher and a PhD student at the University of Ioannina, Greece. After graduating she has been working in schools for several years. She also worked in the field of the Didactics of Natural Sciences in the Department of Early Childhood Education, University of Ioannina where she conducted pilot surveys concerning CHAT and Natural Sciences Education in the early grades.

Eftychia Nanni has bachelor's degrees in Agriculture and Pedagogy and a master's degree in Environmental Studies. She is a PhD student at the University of Ioannina and her research interests include science teacher training and environmental education. She works as a primary school teacher of children aged 6 to 12 years old.

Katerina Plakitsi is associate professor in the department of Early Childhood Education of School of Education at the University of Ioannina, Greece. She is a graduate of Physics and Pedagogy and she has been a school teacher in different grades for 17 years. Since 2003 she has been teaching science education at the University of Ioannina. Her primary researching line is to connect Cultural Historical Activity Theory with science education mainly from the early years and across some formal and informal institutions in national and international level. She is the coordinator of the current reform of the Greek science curriculum.

Efthymis Stamoulis is a primary school teacher and a PhD student at the University of Ioannina in the field of using "History and Philosophy of Science in teaching Science with new technologies." He got a Master's Degree in New Technologies in Education and a second Master in Education. He is co-author of a science web-based multimedia which is used in Greek primary Schools. He also participates to the current reform of the Greek science curriculum.

Xarikleia Theodoraki has graduated as a pre-school teacher. She is a PhD student at the University of Ioannina in Greece. Her PhD Thesis entitled "Designing and analyzing activities of Science Education for pupils aged 5–9 years with the current positions of the Activity Theory (cultural-historical-activity theory – CHAT," is funded by the European Union and national resources in the framework of the program "HERAKLITUS II."

Printed in the United States
By Bookmasters